A MOMENTARY LAPSE OF SIMPLICITY

Creatio ex Nihilo

Ian G. Dalziel

AF808663

A Momentary Lapse of Simplicity, Copyright © 2018 by Ian G. Dalziel. All Rights Reserved.

No part of this book may be reproduced in any form or by any electronic or mechanical means including information storage and retrieval systems, without permission in writing from the author. The only exception is for a reviewer, who may quote short excerpts in a review.

Cover based on *Japanese Zen Garden with Raked Sand* (57820103), by *foodandmore* © *123RF.com*

Ian G. Dalziel
Contact: Info@TheDLPages.com

Produced in the United States of America

First Produced: May, 2018

ISBN-13 978-1-948106-06-1

Other works by Ian G. Dalziel:

NON-FICTION:

The Curious Evolution of Christianity
Prostatectomy: One Patient's Experience

FICTION:

Named for Snow
Princess of the Allevanti
She Dyed for Love
The Kashmir Shrine
Contact in Nevada

CONTENTS

INTRODUCTION

Chaos is the score upon which reality is written.

—Henry Miller

A COUPLE OF YEARS AGO, I wrote a book about religion. It was called *The Curious Evolution of Christianity*, and by writing it I tried to work out why I began to doubt my childhood faith.

The conclusion I came to was that Christianity, like most other belief systems, is a magical fantasy, constructed to gain power over people by playing on humanity's fears and superstitions.

After I had published that work, I sat back, very satisfied with myself. I had done the world a great favor, and now countless millions (gasp, could it be billions?) would buy my book and learn the sordid truth. Those royalties would make me a very rich man.

It was not to be. The millions did not rush to open their wallets, and the penthouse overlooking the Seine remains a dream.

What I then realized was that it is a fine idea to demolish an evil and fraudulent structure. But unless you give people an explanation for the world around them, it is in the nature of humans to just stick with the story they know.

The book you're reading now, *A Momentary Lapse of Simplicity*, aims to provide that missing element. It does so by trying to answer a very difficult question that is often posed to atheists: *If there is no God, then where did everything come from?*

A quick *caveat* for all you *emptores* out there. When I say that I will provide an answer for that tricky question, I am not guaranteeing that I will be correct in doing so. You see, I'm going to have to delve into subjects that I know little of—subjects that I disliked intensely in my early education. Physics, for example.

Who needs physics, I thought as a teenager—only to change my opinion much later in life, as I tried to puzzle out why there is this strange cosmos thingy all around us.

Yet delve I will, unafraid and knowing that a little light brainpower is all you really need. That, and a glass or two of stimulating Cabernet.

We will begin by following the work of scientists through the ages, as they tried to find out what makes up matter. We'll then turn our gaze to the cosmos, again following the trail of those who discovered how our universe functions.

We will look at the origins of life, then will wrap up our journey by tackling that most mysterious of subjects: time. What is time, and does it even exist? If it doesn't, then why do we get old and die?

Every intrepid explorer needs a guide—someone to carry the materials for setting up the camp each night. You will find that I was most fortunate to have a very distinguished companion at my side—a man who, unlike me, knew what he was talking about. Most of the time, anyway.

His name is Albert—Albert Einstein, to be exact.

You're perhaps going to tell me that he died in 1955. That's true, but with a little imagination, I brought him back to life. You should be impressed, since people who do tricks like that usually have churches built for them.

Hmm, the Church of Saint Ian the Divine. I like the sound of that.

There's another warning for my dear readers, namely that science can only take one so far on this journey. Someone once said that when physicists finally discover the truth, they will require a poet to explain it to the world. I'm not a poet, and I know it, but I will do my best for you.

So, please read on, and enjoy this journey with me. I will try to make your reading as enjoyable as possible as we tackle what are usually weighty and serious topics.

A final word of warning may be appropriate at this point. You see, when you get through this material, and make the same realizations that I have about the universe and humanity, you might run screaming from whichever room you happen to be in.

It could be wise to have a pitcher of comforting Margaritas handy.

Pennsylvania, 2018

THE LITTLE MATTER OF MATTER

In which we inspect the small.

MY FRIEND ALBERT

The terrible thing about the quest for truth is that you find it.

—Remy de Gourmont

ONE DAY RECENTLY, I CONFIDENTLY decided that I would learn about and, of course, fully understand this thing called a *ToE.*

Physicists use the acronym *ToE* (it stands for *Theory of Everything*) when they are trying to explain where it all came from—the cosmos, and life in its various flavors.

Let me tell you about the moment I decided to pursue this. I was in Princeton, New Jersey, where I had gone to spend a quiet summer's day.

On that particular afternoon, I wandered round the campus, absorbing what wisdom I could from the cloud of brilliance that hangs above that university. I must say that the beauty of the place made me wonder how one could ever get any studying done in such a lovely and distracting environment.

I popped in to the magnificent chapel. It can sometimes be a little bit gloomy on the inside, although it does have some fine stained-glass windows. On that day, the sun cast beautiful shafts of gentle light into the body of the church.

When I'm strolling around in churches and cathedrals, I'm always slightly concerned that I might burst into flames, or be accosted by the wardens and deacons, or whatever they're called.

On this occasion I daydreamed of just such an encounter.

"Have you anything to confess, my son?" I was asked.

"Let me see," I said, looking at the kindly man who had just approached me, "I suppose I should admit that, while I love visiting churches, I don't believe in a God. Also that I dislike being called 'my son,' which is a sneaky tactic of the Church's to put me in a position of subservience."

"I do represent our eternal and omnipotent Creator, my son," came the confident reply, "therefore I carry within these walls the wisdom of God."

"I see," I replied, "then please help me understand something. Your kind and loving Creator made the universe and all that is in it, right down to the sub-atomic level?"

"That is correct, my son."

"If He did that, then who do you say created Him?" I smiled a cunning smile, knowing that I had trapped this man of God in an impossible spot.

"It's simple, my son. Since God is eternal, he himself has no need of a creator."

"Wow," I said, "you wriggled out of that one quite nicely."

"If you were to attend our services, my son, you would eventually understand."

"That's maybe not such a good idea," I said. "I should probably let you know that I wrote a brilliant and whimsical exposé of the scam that is your religion—in a book called 'The Curious Evolution of Christianity.' Perhaps you've read it?"

"Security!" shouted the man of God. "I need some muscle here!"

I snapped out of my daydream and, unassaulted by that imagined clergyman, moved out into the sunlight, now firmly convinced that I did need to dig into the meaning of everything. Also, that I should describe my thoughts in a best-selling book, for the benefit of all humanity—not to mention the massive royalties

My meandering took me around the campus, then into the lovely town. At a certain point, when I was strolling along Nassau Street, I began to feel a little uneasy, as if I was being followed. I occasionally looked behind me but could see nobody.

I decided that I must be imagining things, so continued on my way, down the incline and into Palmer Square. Ahead of me was the Nassau Inn, where I knew I could have a quick lunch.

Once inside, I ambled downstairs to the Yankee Doodle bar, giving a glance at the framed photos showing some of the Princeton alumni who had gone on to great things.

There was a picture of the actress Brooke Shields, for example. And another of Charles Gibson who hosted *Good Morning, America* for years. There was also a photo of a certain Tom Kean, who was somewhat less successful. He couldn't make it in the entertainment field and had to settle for just being the Governor of New Jersey. His parents must have been so disappointed.

I paused by the bar, to gaze at the original Norman Rockwell painted on the wall. Then I looked at the booths with their wooden benches, and my eye was taken by the odd, disheveled appearance of an elderly gentleman sitting on his own.

I walked over to him.

"Would you mind if a joined you?" I asked.

"Yes, I would," he replied. "I'm trying to have a quiet lunch here."

"I can do quiet," I said, smiling as I slid onto the opposite bench and looked across at the old man.

"What's that you're having?" I said, peering down at his lunch plate.

"A veggie burger with fries," he said. "Over the years, I've tried to go vegetarian, but it hasn't always worked. Look, if that's going to be the extent of your conversation, then you should go."

The waitress arrived, took my order, then walked away.

"Has anyone ever told you," I said, looking at the man's tousled hair and his facial features, "that you look very like Albert Einstein?"

"That's because I *am* Albert Einstein," he said, still bored by my presence. "Although, to avoid freaking people out, I sometimes tell them that I'm made up like this to play the part in a movie."

"But wait," I said, "this is the year 2018. Einstein died in . . . the 1950s, I think. You can't be Einstein."

"How would *you* know?" he said.

"Well," I mumbled, "I mean, it's in the record isn't it? Like, if you look up Wikipedia or something."

"How do you know *that's* correct?" said the man. "Come to think of it, how do you know anything at all?"

I thought for a few seconds. "I dunno. I just think that I know things."

"Pathetic," he said. "When I took my job as a patent clerk, my supervisor told me to consider every single thing written on an application to be false. Even the page numbers. You could do worse than to follow that advice. Tell me, what line of work have you applied your mind to? What do you do in life?"

"I'm a writer," I said, cheering up at the change of subject. I was still a bit disconcerted to be talking to a dead man. I mean, Einstein *is* dead, isn't he? I think he is anyway.

"Have you written anything successful?" he asked.

"Not really," I mumbled. "Right now, I'm working on the origins of the cosmos and of life itself. I need to understand all that for my own satisfaction, but I might also write a book to explain it to people."

There was a long silence. The old man looked at me intently, ate some more of his fries, then said, "I don't think you'll be able to do it."

"Why do you say that?" I asked, feeling offended. "Do you doubt my abilities?"

"Of course. I mean, just look at you, you're obviously a moron." He frowned. "But that's not why. I just don't think anyone can do what you're describing."

"Hmm," I said, "you have a point. It will be difficult to figure out where the cosmos came from, how planets formed and life evolved. Very tricky."

"God, no," he said, "that would be the easy part. It's your other idea that's unworkable."

"The one about explaining things to everybody?"

"Exactly," he said. "There's no way you will ever convince people of the underlying truth. They just won't believe you."

He looked intently at me. "I know I will regret asking this, but who really needs this book of yours?"

"Good question," I said, as I watched the waitress placing my lunch on the table, "and luckily I have an answer. I'm thinking that my book will benefit those who don't believe in a creator. You know, the folks who have abandoned their religion, or never had one in the first place."

"Not clear enough, I'm afraid," Albert said.

"Let me explain," I said, pausing to get my thoughts together. "If someone is born into a religious family—and most people on the planet begin their lives like that—then they grow up hearing a complete explanation for everything around them.

"A super-powerful God made the sky, land and seas. He designed everything, from the planets right down to the sub-atomic thingies that whiz around all over the place. Not just that, but he watches everything, including you and me.

"This scheme is enormously practical, since it relieves you from figuring everything out yourself. These belief systems also contain tricky ways to deal with events that are happy or sad."

"Like what?"

I looked across at him. "If, for example, you are one of the many candidates for a job and you're hired, then your family will tell you that it's God's hand working behind the scenes to help you."

"What about the other candidates?" Albert asked. "Why didn't God help them?"

"Ah," I said, "there will be clever explanations for that. One might be that their faith wasn't strong enough. Another could be that God has other and better plans for them."

"Why wouldn't He have better plans for the one who got the job?"

I smiled. "When the situation gets confusing, you'll be told that everything is part of God's mysterious plan. As a simple human, you're incapable of understanding. That's also what you hear when a relative dies. God needed them in heaven, it's all part of His grand design, and you're not to wonder why."

"You're right," Albert said, reaching across for one of my fries. "That's exactly the crap I heard in Catholic school."

"I thought you were Jewish."

"I was, but I went to Catholic school," Albert said. "I was the only Jew there. And I bought that Christian story for years until I puzzled out the scam. Tell me, how will your book help people?"

I smiled. "If you don't believe in this kind of god, then you are sometimes asked pointed questions, and they're good ones. If God didn't create the universe, then where did it come from? And if there's no God, how was life created?"

"I see," Albert said, "so you'd be trying to show that the cosmos and life evolved without any divine creation whatsoever?"

"I think they call that *abiogenesis*," I said. "When life emerges from non-living material." I leant back on the wall, looked down at my lunch, then picked at the remaining fries on my plate.

"Is everything okay, dear?" said a female voice.

"What?" I said, looking up. It was the young woman again.

"Oh, yes, fine thanks," I said. "The fries are great."

She smiled down at me. "It's just that you've been sitting here for a while, chatting away to yourself. I was hoping that you were okay."

Puzzled, I looked across the table. But there was no old man, and there were no remains of his lunch. Or his glass either.

"Talking to myself?"

"I sometimes do the same thing," said the woman, with a smile. "When I'm caught at it, I just tell people there s nobody else worth listening to."

I laughed nervously. "I must remember that. It might stop folks from thinking I'm insane."

"It's a fine line, isn't it?" she said as she cleared my dishes away. "Who can say whether any of us is a hundred percent, you know?" She paused, then added, "What is it that they say? Everyone is somebody's weirdo?" She chuckled.

"Words to live by," I mumbled, and I took the check, paid for lunch, then went outside into the daylight.

As I walked towards the car park, I wondered about what had just happened. There had been an old man, hadn't there? Surely I didn't imagine that? Because if he hadn't been there, then that would mean I *was* going crazy.

Still, none of that should stop me and my project. Many a lunatic has gone on to write successful books.

Yes, it was definitely time to start my mission. The world is relying on me.

WHAT IS THE SMALLEST THING?

Nothing exists except atoms and empty space.
Everything else is opinion.

—Democritus

I ENTERED THE MULTI-STOREY parking just off the main square in Princeton and found my car. I unlocked the driver's door, opened it, then settled down in the comforting genuine faux leather.

"Hello again," said a male voice.

I turned to my right and saw . . . Albert, the old man from the restaurant.

"I don't understand this," I said. "You don't really exist, do you?"

"They always say that," he said. "Drives me mad, and I'm only trying to help." He paused, then said, "You see, I've decided to take you in hand."

"What does that mean?"

"I'm going to give you a guided tour of some of the key events in science. We'll start by following the history of man's search to understand what makes up matter."

I shook my head. "That's very kind of you, but I think I can manage on my own."

"Clearly untrue," Albert said. "While you walked here, I checked into your history as an author. Turns out nobody's buying your book on Christianity. Or any of your other works, like those tacky little novels." He sniffed.

"Well," I said, feeling guilty, "it takes time to build a following."

"In the meantime," he said, "let's begin our journey. I'm going to take you back to ancient Greece, to meet a couple of men called Leucippus and Democritus."

"You can do that?"

"Of course," he said. "I'm a Time Lord. I can take us anywhere. And anywhen, for that matter."

"You mean like that Doctor Who character on television?" I asked.

"Yes," Albert said. "Except that he's fictional, and this is real."

Suddenly, a mist occupied the interior of the car. After it cleared, I found myself hovering in the air, looking at a young man hunched over a table. He was working outdoors, and the sunlight around him was extremely strong.

"It's around 400 BCE, and that man's Democritus," Albert said, as he hovered beside me. "Look, here comes his master, Leucippus. These two may have been the first to lay down a formal theory of matter, namely that everything is made up of small, indivisible particles.

"Their idea being that if you take something like a piece of wood or stone, break it in two, then keep splitting again and again, you will get to a point where you can't divide it any more. You'll then be dealing with some kind of fundamental particle. Let's listen in."

"How far along are you, Democritus?" Leucippus said, leaning over his pupil's shoulder.

"Hey," I said to Albert, "how come I can understand them? Shouldn't they be talking in Greek or something?"

"Translation is all part of the service," he replied. "No extra charge. Just listen."

"It's getting really hard to split the wood further, Master," said the young man, putting down his knife. "Can't we just assume that there would be a limit to the splitting?"

"Assume?" snorted Leucippus. "You see, this is why I'm the master, and you're just a pupil that nobody will ever remember. Now keep going till your fingers bleed."

"Well, then," Democritus said, "can't you at least install a particle accelerator here, so I can get some meaningful work done?"

"Particle accelerator?" said Leucippus, with a puzzled frown. "What in the name of Hades, son of Cronus and Rhea, is that?"

"I dunno," Democritus said, frowning as he looked up at the sky. "The words just popped into my head. It's probably nothing. Forget it."

He picked up his knife, bent over the workbench and started to slice again at the tiny pieces of wood on the table.

Albert chuckled. "In spite of what Leucippus said, it was Democritus who turned out to be the star of the duo, and he went on to develop some exciting theories. He felt, for instance, that all matter was made up of tiny atoms.

"Hard substances, he thought, had atoms that gripped each other strongly, perhaps with tiny hooks or barbs. Softer atoms, like those in liquids, were more slippery."

"Not bad," I said, admiring the intelligence of the ancient men.

"Not everyone was a fan," Albert said. "Plato disliked Democritus and his ideas so much that he wanted the man's writings burned."

I frowned. "I used to think well of Plato. I might have to change my opinion." I paused, then said, "Wait, didn't they talk about things being made of earth and fire back then?"

"They did," Albert said. "Many old-timers felt that matter was made up of just four so-called elements: earth, fire, air and water. Plus, for some folks, like Aristotle, there was a fifth element that represented the void or spirit beyond the physical world. That idea persisted for many centuries, as did the term *quintessential*, for that mysterious fifth element.

"Now, let's fast-forward into the seventeenth century and we can meet Robert Boyle. He was born in Ireland and was the son of a very wealthy Earl."

The mist swirled once more and, when it cleared, I couldn't quite tell where I was.

It didn't look like Ireland at all, at all.

BOYLE'S LITTLE THINGS

Those distinct substances, which concretes generally either afford or are made up of, may, without very much inconvenience be called the elements or principles of them.

—Robert Boyle

WHEN YOU FINISH YOUR EDUCATION at Eton, you will have acquired a little bit of knowledge and, more importantly, an upper-class English accent. Ex-Etonians then usually go on to be Prime Minister or something like that. They don't really want to, but it's expected of them.

I didn't go to Eton myself, because I don't care for politics. (Yes, *that* was the reason. It wasn't that I wasn't clever or rich enough. Of course not.)

Albert spoke quietly. "Robert Boyle did go to Eton, but he managed to dodge the political life, preferring to put on a white smock and mess around in his lab. One of the wonderful things he did was to spend a winter in Florence with Galileo, swapping exciting ideas."

"Is that where we are now?" I asked, looking at the hills covered in vines. "That would explain why the grass isn't green like in Ireland."

"That's right," Albert said. "This is Italy, in 1641. Let's hear what's on their brilliant minds."

"Whose turn is it to grill the funghi porcini?" asked Boyle, looking across the room at Galileo.

"I did it yesterday," said Galileo. He paused, then said, "By the way, Bobby, how long are you going to stay here, eating my food and drinking my wine?"

"Cool it, Gally," said Boyle, "don't get your leggings in a twist. I'll nip over to the vineyard and buy a flask of rosso. And if you don't mind, I'd like to stay here until you've explained that glass thingy you've been working on. What did you call it? A telescon?"

"Telescope," said Galileo, "and I've just made a new model that brings in distant objects with much better clarity. Soon I'll be able to see cargo ships from many miles away, long before they reach port. If I can recognize the vessel, I'll know what it's carrying, and I can place some quick bets on the commodities market." He smiled. "I'm going to make a killing."

Boyle looked thoughtful. "I suppose," he said, "but don't you think it would be more beneficial if you used the telescope on everyday materials, like sand or water?"

"Why on earth would I want to do that?" mumbled Galileo, as he peered out of the window and fiddled with the telescope's focusing ring.

Boyle smiled. "You could maybe see what matter is made of. You know, look closely at those little atomies the ancient Greeks talked about."

"Next thing," said Galileo, with a sneer, "you'll be wanting me to look up at the moon and see if it's really made of cheese." He frowned. "Cheese!" he said excitedly, taking his gaze from the telescope to glower at Boyle, "that's another thing we need. You finished the last of my Gorgonzola yesterday."

"Robert came back home," Albert said, "and around 1661, he theorized that matter was made up of substances that could not be broken down."

"Hang on," I said. "Around 400 BCE, Democritus said matter was made up of indivisible particles? And it wasn't till 1661 that Boyle picked that up again? That's about two thousand years."

Albert nodded. "It's true, progress was agonizingly slow during that long, dark period. Fortunately, Boyle marked the beginning of a new era for us, and he did more than just talk about these things. He developed methods for analyzing compounds, and many think of him as the earliest modern chemist.

"He also had a wish list, where he defined the top advances he'd like to see in civilization—things like the ability to fly, and the invention of drugs to alter the imagination."

"Not bad," I said. "If the little hippie had hung on a few more centuries, he'd have shared in the misery that the aviation industry inflicts on us all."

Albert chuckled. "I doubt it. With his brilliant mind, Boyle would probably have wangled an upgrade to first class."

Albert looked intently at me. "You were just complaining that it took so long to make progress on the elements? Luckily, Boyle and his contemporaries really started to move things along. For example, there was a German businessman, Hennig Brand, who went bankrupt, then tried to discover a process that would turn ordinary metals into gold."

I chuckled. "I try to do a similar thing myself, by turning the ordinary paper of a lottery ticket into hundreds of millions of dollars. Haven't succeeded yet, but others have done it before and I'm just trying to confirm their research."

Albert smiled wryly. "It's surely a loss to the world of physics that you decided to become an author instead. Moving along . . . around 1669, our friend Hennig was messing around with distillation techniques, using urine in his experiments."

"That's because it was free and plentiful," I said. "If anyone wants mine, they can just give me a call."

"Hennig ended up with a white substance that he named *phosphorus*," Albert said. "He probably didn't know this, but you don't easily find pure phosphorus anywhere, even though it is important for life. It is in our DNA, after all.

"Hennig didn't think the phosphorus thing was all that special, so didn't publish his findings till Bobby Boyle repeated the work. Those combined efforts got folks excited about what the basic elements really were.

"A contemporary of theirs was a French researcher called Nicolas Lémery, who joined the discussion. He was born in 1645 and was a Calvinist, so he had to flee from Catholic France to England. He found his way back to France a few years later, when he converted to Catholicism."

"No surprise there," I said. "Who wouldn't convert to Catholicism if it meant a daily supply of baguettes and runny Camembert?" I looked at Albert. "I say, why aren't we visiting all those researchers you're describing?"

"We can't do that with all of them," Albert said. "We don't have time. Or, rather, you don't. I have all the time in the world."

"Point taken," I said.

Albert continued. "As well as being a theory nut, Nicolas enjoyed experimenting. He messed around with acids and alkalis, trying to figure out how their elemental particles combined. For example, he thought acids might have little spikes on them while alkalis had pores or dimples like a golf ball. Frenchie Nicolas, however, is remembered because he may have been one of the first to think deeply about how elements might fit together, like interlocking puzzle pieces. And then, Amadeo was born."

"The composer?" I asked. "Mozart did science stuff?"

"No," Albert said, "Mozart's name was Amadeus, not Amadeo. It's really alarming how your brain muddles things up. It's just as well I decided to step in and keep you on track.

"Amadeo *Avogadro*, in fact, was born in Turin in 1776. He practiced ecclesiastical law, before getting the science bug and setting up his test tubes. In his lab, he did important work on gases, and there's a scientific law bearing his name. More importantly for us, he tried to resolve the confusion around atoms and molecules, since people were using the terms interchangeably.

"He stated that gases were made up of *molecules*, and that molecules were made up of *atoms*. He published papers describing his various theories, but they were largely ignored. It wasn't till a few years after Amadeo died, that his work was accepted."

"Brilliant," I said. "I'm assuming he won a Nobel for his efforts?"

"You don't know much, do you?" Albert said. "The Nobels are not given to dead people."

"I see," I said. "Lucky for me then, that I'm living in an age when Nobels *are* being handed out." I turned and smiled at Albert. "I'll probably win one for the book I'm going to write."

"Of course you will," Albert said, dryly. "Now, let's visit one of the all-time greats—a man called Mendeleev."

"My goodness," I said, "imagine seeing him. We're not worthy."

"You're not, but I certainly am," Albert said. "I'll get the mist going again."

MENDELEEV'S PATTERNS

For the first time I saw a medley of haphazard facts fall into line and order. All the jumbles and recipes and hotchpotch of the inorganic chemistry of my boyhood seemed to fit into the scheme before my eyes—as though one were standing beside a jungle and it suddenly transformed itself into a Dutch garden.

—C.P. Snow

"YOU MENTIONED MOZART A WHILE BACK?" Albert said. "Well, back in 1766, when he was already a famous ten-year old piano prodigy, a very talented young boy was born in Cumberland, the most north-western county in England.

"People still rave about Wolfgang and his catchy musical numbers, but hardly anyone gives a thought to the young English Quaker boy called John Dalton. Yet he made arguably more of an impact on our lives through his work on atoms, than Wolfy ever did with his music."

"I find that hard to believe," I said. "Have you ever heard Mozart's Requiem?"

"Of course," Albert said. "I was a violinist myself, and Mozart was one of my favorite composers. In fact, if I hadn't been so busy working on relativity, I'd have been a musician of some kind. I did actually perform in some fund-raising concerts, you know."

"What did Dalton do?"

"John was born into a poor family and was educated by his father and another Quaker, but the boy must have been clever since, by the time he was twelve, he was earning a living teaching other children. Remarkably, by age fifteen and aided only by his brother, he was running a Quaker school.

"John then bounced around for a while, educating youngsters while pursuing his interests in several fields. He wrote papers on mathematical and meteorological issues, and then became well-known for tackling the subject of color blindness."

"Why that in particular?" I asked.

"Both John and his brother were color blind, so John suspected it must be a hereditary condition, and that fascinated him. When he died, his eyeball was preserved, and in the 1990s scientists were able to tell that John could see only one color, yellow. For his work on eyes, he was honored by having color blindness called *Daltonism*."

"I've never heard it called that," I said.

"It's no longer the general term for color blindness," Albert said. "Dalton was also important because he was intrigued by what happened to substances as they were subjected to pressure, boiling and freezing. As he experimented, he made careful measurements that he recorded in his notebooks.

"He somehow came to the conclusion that differences in how substances reacted were due to the weights and particles of their underlying components."

"How did he figure that out?"

"Scholars can't tell," Albert said. "They have examined John's documents to see what might have influenced him, but it seems that an independent insight helped Dalton to come up with his ideas.

"He said, for example, that elements were made up of small atoms, and that a particular element's atoms would all be identical. He also said that atoms were indivisible and could not be destroyed. Neither of those ideas is totally correct."

"But good enough for government work, right?" I said, trying to lighten the mood.

Albert let out a reluctant chuckle. "John also went on to state that atoms could be joined through chemical processes to form compounds. And, yes, he was right about that.

"Perhaps his most important contribution, however, was that he published a list of around half a dozen elements, giving each a value that corresponded to what we now call their atomic weights. That inspired others to expand his early set, and by 1828 a Swedish Baron called Jacob Berzelius had produced an expanded list of elements.

"It is, of course, one thing to measure, record, and list attributes. It's entirely another to see structure behind data. For that, we had to wait for the big dog in the atomic weight arena, a certain Dmitri Mendeleev."

"A Russian?" I asked.

Albert nodded. "He was born in 1834, in a small village in Siberia. His mother was intent on the boy being educated, so she took him all the way to Moscow in the hope of enrolling him in the university there. Sadly, they were turned away but, undeterred, she found him a place in St. Petersburg."

"I prefer St. Pete anyway," I said. "Nicer palaces."

"Dmitri must have been a clever chap," Albert said, "since he received a doctorate for his work on alcohol and water."

"I sometimes think," I said, "that I ought to have recognition for my own work on alcohol. I study the composition of the Martini, every afternoon."

"You know, you can be quite tiresome," Albert said. He paused, then said, "By the time Dmitri got excited about elements, scientists were discovering a new one every year. He became intrigued, wondering how on earth all these elements fit together.

"As even you probably know, he figured this out so well that he is known to us as the father of the Periodic Table, that huge poster that's hung in schools, and that most children gaze at in confusion.

"There is, however, a rather intriguing thing or two to know about how Dmitri did his work, and I wonder what we might overhear if we go there and look over his shoulder."

Mist started to swirl around us.

"Oh, does this mean that Mendeleev was important? If he rates mist, I mean?"

"Very important," Albert said. "Just listen and learn."

"Whatcha doing, Dmitri?" asked a tall man as he approached the table where Mendeleev was sitting. "Playing Patience?"

"Patience, Antonin?" Dmitri said, looking up at his colleague. "What's that about?"

"You know . . . Patience," Antonin said. "It's a card game you play to pass the time." He paused, then looked down at the table. "Oh, sorry, those aren't playing cards, are they?"

"They're cards I made for the elements," Dmitri said, his gaze returning to the pattern on the table. "I'm trying to arrange them in some kind of system." He smiled. "I got the idea in a dream I had."

"Is a structure even possible?" Antonin asked. "I mean, I'd be surprised if the elements followed any kind of rules. Aren't they just the chaotic stuff of the universe?"

"I'm not so sure," Dmitri said.

Antonin gazed at the cards. "Oh well, I can see a couple of problems right away. You've put down the wrong atomic weights for some entries, and I really don't understand why you've arranged your rows and columns the way you have."

"My vertical columns," Dmitri said, his voice calm, "are for substances that are similar. For example, I've put these elements together in this column, because they're all soft and easy to cut, among other things."

"You've also left spaces," Antonin said. "You can't be doing that if you're trying to build a complete pattern."

Dmitri sighed. "Those gaps are where I think there may be elements we haven't yet found."

"It's intriguing," Albert said. "If Dmitri had used only the elements and atomic weights that were considered correct in his time, he would probably never have found the pattern that gave us the Periodic Table.

"Instead, he used his judgment, changed weights where he felt it necessary, and left gaps where he thought there were elements to be discovered. Scholars who have examined his work have no idea why Dmitri did this the way he did. It all seemed to be based on just a hunch."

"Interesting," I said. "So, our understanding of the elements was pushed forward by these two men, Dalton and Mendeleev, each of whom did more than just conduct experiments and record results. They used their intuition in spectacular ways. You know, I feel a kindred spirit with them, since I share their tremendous talent for discovery."

"Of course you do," sighed Albert. "Now, at this point in history, they were making great progress. But you'll never guess what they started to argue about next."

I frowned. "Atomic structure?"

"Nope," Albert said. "They got into an argument about what to call themselves."

WHAT'S IN A NAME?

I have a poor memory for names, but I seldom remember a face.

—W.C. Fields

AFTER THOSE EARLY BREAKTHROUGHS, things started to move very quickly, and the sales of beakers, burners and lab benches soared."

Albert was really getting into telling his story now and was speaking with enthusiasm.

"With everybody messing around in their basements, a most important thing for them to do seemed to be . . . figuring out what to call themselves."

"I don't understand," I said.

"Up to that time," Albert said, "they had been using the term *natural philosophers*, but that clearly wouldn't do—it took far too long to write on their business cards. So, a certain William Whewell, an English thinker, seems to have started using the word *scientist* in the nineteenth century. Whewell was a regular speaker at meetings in Cambridge University.

"You see," Albert went on, "many researchers preferred not to separate the different fields. You'd see books with titles like *The Philosophy of Plants.* But there came a point when many wanted to adopt specific names for different branches of thinking."

"Whewell made the change?" I asked.

"Partly as a joke," Albert replied. "He suggested the word *scientist* from the old French word *science*, meaning knowledge. He was reacting to some of the self-described philosophers who didn't like associating with people at lab benches, surrounded by flasks of smelly liquids." Albert laughed. "Samuel Taylor Coleridge was one of the whiners."

"The poet?"

"Not just a poet, I must say," Albert said. "He was also a decent thinker, but unfortunately a vain man." Albert laughed. "It's said that he was once in his garden, reading aloud one of his poems to his friend, Charles Lamb. To make sure Lamb wouldn't escape, Coleridge grabbed hold of his friend's coat button and held on."

"What happened next?"

"Lamb took out his penknife and cut off his own button. A short while later, when he was returning past the garden, he looked in and saw Coleridge still standing there, eyes shut while dreamily reciting the poem and . . . still hanging on to the button."

"You think more highly of Whewell?"

Albert smiled. "Whewell seems to have been a clever sort, and a bit of a wordsmith. He's credited with several new terms in his field—like *anode*, *cathode*, and *physicist*. Not bad for a dirty-sleeved lab worker."

"But them's just words," I said. "I don't see how that helps us much."

"There is a point to my story," Albert said. "In olden times, a thinker could reasonably expect to have a good understanding of all branches of science. Eventually, each field collected so much knowledge, and became so specialized, that you could no longer expect to know everything. In my own case, I had to take time out to learn a whole new branch of mathematics to pursue my theories."

"Which branch?"

"Riemannian geometry," Albert said. Seeing my blank stare, he smiled. "It deals with the mathematics of curved objects."

"I see."

"I wasn't the only one delayed in my research," Albert said. "Those pursuing quantum theory, like Born and Heisenberg, had to turn to a man named Pascual Jordan to learn how to multiply matrices. The math for that is also different, you see."

"You said you had a point in all this?" I said.

"Yes. There are so many fields of knowledge to absorb, so how on earth are you, unschooled in any of them, ever going to figure things out?"

"Don't misunderestimate me," I said.

"Misunderestimate?" Albert said.

"You know what I mean," I said. "And need I remind you that I don't need to prove my conclusions to the satisfaction of the scientific community? I just need to satisfy myself."

"Hmm," Albert said. "You know, I could leave you to flounder, but I feel a duty to the world of physics. I can't have you producing a book that's full of inexact thinking. Let's continue our journey. Our next step will be to visit someone who did a wonderful thing. He invented X-rays."

"Wonderful enough for mist?" I asked.

"Most definitely mist-worthy," Albert said, as it started to swirl.

"Oh good, I like the mist thing," I said.

"I bet you like sprinkles on your ice cream too, don't you?"

RÖNTGEN'S X FACTOR

I have also a shadow of the bones of the hand, of a wire wound upon a bobbin, of a set of weights in a box, of a compass card and needle completely enclosed in a metal case, of a piece of metal where the X-rays show the want of homogeneity, and of other things.

—Wilhelm Röntgen

"THIS LOOKS LIKE SOME KIND OF old-fashioned workshop," I said as the mist cleared. "Where is this?'

"It's in Germany," Albert said.

"There's nobody here," I said.

"Have patience. We're just a tad early," Albert said. "Time travel is only accurate to within a minute or so."

Just then, the door of the workshop opened, and two men entered.

"The one who came in first is Röntgen," Albert said. "Around 1895, he discovered something he was calling X-radiation."

"Why X?"

"Because X was the usual symbol for an unknown factor," Albert said. "Röntgen was fascinated when he found out that these magical rays could see through materials to pick out objects that lay behind them. He wanted to know what the rays were, and at first thought they might be a particular form of light. You see Röntgen knew that prisms could deflect light, so he tried that on the rays."

"What happened?"

"Nothing," Albert said. "And that was a pity, since Röntgen expected that focusing X-rays might cause them to do something really spectacular, just as focused light can burn objects." Albert smiled. "Kids with magnifying glasses still delight in finding this out, as they gleefully focus sunlight to burn paper. Or occasionally burn their parents' houses to the ground, the scamps."

"What was he doing with the rays?"

"Röntgen had already used the rays to see through material to produce images of objects like watches, nails, bones, and so on. That meant there was a series of excited visitors coming to check things out in Röntgen's lab. Like that man who followed him in just now."

"Professor Röntgen, is this true?" said the bearded visitor. "These X-rays, as you call them, they can see through dense matter and produce images of what's inside?"

"Indeed so, Professor Stein," said Röntgen. "I have colleagues who like to play with what are called Crookes tubes. They're evacuated glass containers that have two plates inside. It was found that, if you pass an electrical current through the vacuum, then a spark is produced.

"But that's not all. I decided to coat the metal plates with barium, then paint the outside of the tube black, forcing the beam to escape through a small patch of foil. I put a barium screen close by, then placed a book in between."

"Why, dear Professor?"

Röntgen smiled. "To see what would happen, of course. As it turned out, in the pages there was a key I was using as a bookmark, and I had forgotten all about it. To my amazement, the barium screen showed an image of the key!"

"You found out how to see through things," Stein said, his tone enthusiastic.

"Exactly," Röntgen said. "Here, let me show you how this works."

He moved to the center of his laboratory, then readied his equipment.

"I have positioned a sensitive plate in the test area here. Now, why don't you choose an object from the arrangement on the side table. There are nails, bullets, bones, and so on. When you've picked one, you can lay it on the plate and we'll cover it with a dense material."

"Could I use my own watch?" Stein asked.

"Ah," Röntgen said, smiling at his friend, "I can see you are a cautious man, Professor. You have perhaps been listening to Lord Kelvin, who said that X-rays are nothing but a fraud? You think I might have set up pre-prepared images for all those objects on the table? I admire your scientific discipline. Yes, of course, please place your watch right there."

"It won't be damaged, I hope?"

"Nope," Röntgen said, "not a scratch. And if it's working now, it'll be working just fine afterwards."

The professor took his pocket watch out, then placed it on the test bed.

Röntgen looked down at the watch. "It's a fine one, isn't it? Nicely engraved."

"It was my father's," Stein said. "The inscription shows it was a gift to him from my mother."

Röntgen placed a thin sheet of paper over the watch, then placed a wrapped package in Stein's hands.

"Beef," said Röntgen, with a chuckle. "For the dogs' dinner later. Please place the package over the watch."

Stein did as requested. "The dogs can eat the beef after it's been through this process?"

"They seem to enjoy it," Röntgen said. "Now," he said, "we have to take refuge behind the screens over there."

"Why?" asked Stein. "Are the rays dangerous to humans?"

"I don't know," Röntgen said, "but I'm not taking any chances."

The two men sheltered behind lead-lined screens, then Röntgen activated the X-rays.

A second later, Röntgen stepped out from behind the screens, and walked back towards the test equipment.

"That's it?" asked Stein. "I was expecting something more dramatic, but I heard almost nothing."

"Sorry to disappoint, my dear friend," said Röntgen. He handed Stein back his watch. "All that remains is to prepare and inspect the image."

"It's fantastic," said Stein, later. "It shows the case, the hands and the numerals perfectly. Even the engraving is visible. And all that done through solid flesh."

"Indeed," Röntgen said. "Just think of the benefits this will bring to doctors and dentists."

Albert chuckled. "Röntgen was right. As you might imagine, everyone immediately cottoned on to the medical and dental benefits of these new-fangled X-rays. They used them to locate bullets inside bodies, to see whether broken bones had been correctly reset, and to examine teeth."

"But we know the rays are dangerous, don't we?" I said.

"We do," Albert said. "It was eventually discovered that there were problems for those working with the new X-rays. After a while, some technicians had burns on their skin and damage to their eyelids and lips. There were even reports of collapse and death.

"Röntgen had been wise to exert caution and to shield himself with lead. That helped him live long enough to receive a well-deserved Nobel Prize for Physics. The very first one, as it happens."

Mist began to swirl again.

"Someone important coming on deck?" I said.

"Absolutely," Albert said. "A diva, if the truth be known."

When the mist cleared, I smiled at Albert. "I know where this is. It's one of my favorite cities, Paris."

"*Mais oui, mon cher*," Albert said.

CAN ATOMS DIVIDE?

One never notices what has been done.
One can only see what remains to be done.

—Marie Curie

I LOOKED DOWN AT THE SCENE where a young woman was sitting across the table from a much older man.

"Who's she?" I asked Albert.

"Her name is Sklodowska," replied Albert.

"Never heard of her," I said.

"You have," Albert said. "Just be patient."

There was silence for a few seconds, then Albert said, "This young woman is from Warsaw, and she's nervous because this is a job interview. She's been in Paris for a while now, doesn't have much money, and can only afford to rent a tiny attic.

"Her life was different back home in Poland. Her father had been a scientist who had kept lab equipment at home, where she enjoyed helping him with his work. But he fell on hard times and now here she is, alone in Paris in 1891. A sad tale."

"Mademoiselle Sklodowska, is it?" said the man, as he looked across the table at young woman.

"That's right," Marie said. "You received the letter I sent?"

"I did," said the man. "You're from Warsaw, it seems. And you say you have some experience in tutoring? I need someone to help my son."

"I would often help my father," she said. "I'd clean his test equipment, help him set up his lab experiments, and generally tidy up when he was done."

The interview continued for some time, and the man looked closely at Marie as she answered questions in her calm way.

Her appearance was neat, she was soberly dressed, and she looked and sounded as if she was a no-nonsense type. Maybe she'd fit in well with his son.

Her French was good, too. She had an accent, of course, and that was to be expected; this wasn't her mother tongue, after all. But her grammar was perfect, and her vocabulary good. She even used the subjunctive correctly. Yes, this woman might be a good teacher.

"Very well, Mademoiselle," the man said, "you may start next week."

"She got the job," I said, with a smile.

"Indeed," Albert said. "Her story turned interesting rather quickly. She worked hard at her own studies and earned two degrees, then went on to work in an industrial laboratory, where she was studying the composition of steel. A Polish physicist knew that Marie was looking for a larger laboratory space for her work, so he introduced her to a scientist called Pierre Curie."

"Oh," I said, "she's Marie Curie. I see now."

"It took you long enough," Albert said. "Pierre kindly rearranged his working areas to give Marie the extra room she needed. He was also impressed by how diligent she was.

"You know, we tend to think of Marie as having been somewhat stiff and sober. There is, after all, that story about an assistant who rushed up to her, alarmed at having just accidentally swallowed a pin. 'Never mind,' said Marie, 'here's another pin you can use.'"

Albert chuckled. "But when it came to Pierre and Marie, there must have been some monkey business going on, and the two of them fell in love. Marie then made things problematic by insisting that she really wanted to return to Warsaw. She went there for a summer break and was happy to be reunited with her family.

"When Pierre wrote to say that he'd drop everything and follow her, she realized that it must be true love, so she returned to Paris and they got married. For years afterwards, Marie, practical as ever, wore her dark blue bridal outfit as she worked at her lab bench. That's where she got involved with the element uranium, and that's what she and her hubby are known for."

"Yet another happy ending," I said.

"Just wait a bit," Albert said. "There's sadness too."

"How did she begin working with uranium?" I asked.

Albert nodded. "There had been a lot of interest in the X-ray work done by Röntgen. A researcher, Henri Becquerel had discovered that uranium salts emitted rays that were similar. Uranium also seemed to be warmer than surrounding materials, without receiving heat from any other agent.

"Marie didn't particularly care about uranium but wondered whether it might make a good topic for a thesis. You see, Pierre and his brother had invented a device for measuring electric charges, and Marie employed it to see what was going on with the element.

"She found that the atmosphere around uranium samples was electrically charged. Marie concluded that this was not because something was acting on the uranium; it was the element itself that was emitting particles."

"That's major," I said. "Because up till then, they had all thought that atoms couldn't be broken up."

"So you do have a functioning brain," Albert said. "You're right. The world was most excited about this discovery. When Pierre gave a lecture on their findings, the hall was packed, and everyone watched in awe when the lights were dimmed to show that the element emitted a slight glow.

"A small amount was then displayed on a Paris avenue, and people walking to work would stop and gaze in wonder at this exciting material."

"Different times," I said. "I can't imagine anyone stopping on Fifth Avenue to look at a new element. A new iPad, maybe."

Albert laughed. "Marie went on to show that thorium does a similar trick. She missed getting credit for that work by just a couple of months, because a competing scientist, Gerhard Schmidt, managed to publish first.

"Marie never forgot that, and thereafter made sure to release her results as quickly as possible. She and Pierre went on to discover *polonium*, named after her native land, and *radium*."

"Marie must have been famous by then," I said.

"She was," replied Albert, "but she ran into some issues with the scientific elite. When the Royal Institution in London wanted to hear about the Curies' work, they made it clear that only Pierre would be allowed to address their assembly. Good heavens, they mumbled among themselves, we can't allow a woman in here, can we?"

"Outrageous idea," I said. "Bar the doors, chaps. Pass the brandy and cigars."

Albert laughed. "In 1903," he said, "Pierre and Marie shared the Nobel Prize for Physics with Henri Becquerel. She nearly didn't receive the honor, but Pierre was alerted by someone who noticed that Marie had been quietly dropped from the award list.

"Pierre managed to get her reinstated, and that is how Marie, the young woman who had come to Paris, had lived in an attic eating only bread and milk, and whose early work had been ridiculed by her peers, became the first woman to win a Nobel."

I smiled. "Well done, Marie."

"I agree," Albert said. "She won a second Nobel in 1911, for Chemistry. But as I said earlier, there was sadness. In 1906, husband Pierre was struck by a carriage in the street and died of his injuries.

"There was simply nobody else around with the knowledge to continue Pierre's research work. It could only be Marie, and that was a matter of great concern to the mandarins in the Sorbonne. They'd never before appointed a woman as a professor but, to their credit, they gave Marie the nod."

I smiled. "Thanks for giving me a glimpse into her life. It's nice to see how things were quieter back then. There was no internet, no TV screaming in your face all day long. People worked calmly at their lab benches, in a more unhurried and civilized way."

"Don't you believe it," Albert said. "Then, as now, scientists like Marie worked in a fury, pulling all-nighters to be the first to find a new element or make a new realization. That intensive, non-stop work done by the Curies had implications that lasted long after their passing.

"Marie paid a price for success, however, since her own death was caused by the very substance that gave her fame. She had carried radioactive samples around in her pockets, and had the material close to her workbench all the time. To this day, her research papers are too radioactive for anyone to handle."

"But they had shown that atoms could break down," I said. "That also meant that atoms must have an internal structure of some kind, made up of still smaller particles."

"Right," Albert said, "and this new idea immediately sent white-coated boffins rushing to their lab benches, eager to figure it out. Bit by bit, they did. Ernest Rutherford summed it up quite nicely for everyone. Let's meet Ernie."

A TINY SOLAR SYSTEM

All of physics is either impossible or trivial. It is impossible until you understand it, and then it becomes trivial.

—Sir Ernest Rutherford

I DID LIKE THE NINETEEN-TWENTIES, you know," Albert said. "The Roaring Twenties, they called them. With World War One behind us, everyone was upbeat, and progress was being made in all sorts of interesting ways. The stock market collapse of 1929 would shatter that mood, but that was still a few years away."

"What was life like back then?" I asked.

"It was exciting," Albert said. "A most important change, in my view at least, was in women's clothing. This was the era of the flapper girls, you see. Carefree young women began to dress differently, in ways that would give their corseted grandmothers tremors, and send those dear old things to their fainting couches."

"Yes," I said, "I've seen photos of flappers. Fringy skirts and cloche hats. I love that look."

Albert continued. "It's not certain where the word flapper came from, but some say it was from the image of a girl wearing long hair in a pony tail flapping on her back as she skipped along. Such a nice mental image of a carefree young thing.

"The French were partly to blame, or perhaps thank, for the new style of dress: light dresses with spaghetti-thin shoulder straps, and fringed hems that showed more leg than had been seen before. Lustful French men had also come to prefer women who were slim, even flat-chested, a style that was known as *garçonne*, meaning boyish."

"I'm not exactly against that myself," I said.

Albert nodded. "It's unlikely that Sir Ernest Rutherford paid much attention to things like that. He was, after all, trying to figure out the universe and all its workings. Here he is, getting ready to summarize all the exciting breakthroughs in understanding the atom."

The young assistant looked nervously at the great man, "Have you finalized your speech yet, Sir?"

"Not quite, William," the distinguished scientist said. "There is a lot to cover, you know." He paused, smiled at the young man, then said, "I have to make sure I get my message across to the lowest denominator. That makes it tricky."

"I don't understand," said William.

"It's all so new for them, you see," said Sir Ernest. "For thousands of years, mankind has thought that matter was made up of a few immutable components."

"You mean, like . . . earth, fire, air, and water," said William, nodding.

"Quite," Rutherford said. "Of course, Mendeleev and his ilk helped us greatly by defining a huge list of elements. He went on to organize them into an intelligent structure. That was wonderful work, and I can hardly believe how much progress we've made in just a couple of decades. I'm also thinking of the Curies' discoveries with uranium. They were able to show that atoms are actually fragmenting all the time, and that was a colossal breakthrough."

William nodded. "Your own work wasn't too bad either, Sir," he said. "Figuring out the structure of the atom itself."

"Ah, the atom," sighed Albert. "Rutherford was one of those who favored the idea that the atom of any element was like a tiny solar system, with the nucleus holding some electrons tightly, with others buzzing around in circular orbits, perhaps at speeds of one hundred thousand miles per second.

"He saw all that as rather appealing. After all, our solar system has the sun with its orbiting planets. What could be more intellectually satisfying than seeing this simple design reflected at the lowest level of matter?"

"What specifically did Ernie do?" I asked. "Did he discover anything?"

Albert nodded. "Just as the Curies had done, Ernest spent a considerable time working with uranium. He thought that the heat issuing forth was in the form of what he called alpha particles, perhaps as many as one hundred million of them emanating every second from just one milligram of the element.

"He figured that those particles occasionally hit the core of a target atom, and from that he concluded that the atom's core must be relatively dense. He suspected that most of an atom is empty space, although he did wonder what stopped the orbiting electrons from simply crashing down into the nucleus.

"He also thought that one element could be transformed into another if the correct process could be found."

"You know, William," said Rutherford, "just think back to how the ancient Greeks noticed that rubbing amber with fur or cloth made it attract other materials. We then used their word for amber, ēlektron, as the root for our own words electric and electricity.

"Then we had to wait until the 1700s for Hauksbee and De Fay to begin figuring out how electricity worked. They theorized that there were different types of electricity, until a certain Benjamin Franklin corrected them."

"A brilliant man, Franklin," William said, "even if he was an American."

"It's astonishing, isn't it?" Ernest said, nodding. "Some of them can be quite clever."

He continued. "In the early 1800s, Laming suspected that the atom was a core, surrounded by electrically-charged particles. Smart man, Laming.

"By the end of the 1800s, they were all arguing about the correct word to use for these particles. Some wanted electrolion, others wanted electrion, but eventually we all settled on electron."

"Well," William said, "if we've all managed to agree on a name for the particle, then that is startling progress indeed."

Ernest laughed. "Couldn't have put it better myself, William. Do you mind if I use that thought in my speech?"

"Another thing to consider adding, Sir," William said. "You could inject some humor by saying that you've banished plum pudding from the menus in science labs around the world?"

"Plum pudding?" Rutherford frowned. Then he laughed, 'Of course, you mean that I have killed off J.J Thomson's design for the atom—like a pudding with all the electrons stuck firmly in the nucleus."

"Exactly, Sir," William said. "The news of your design will shake the scientific world. And its canteens."

"Did it?" I asked, looking at Albert.

"Not in fact," he replied. "Hardly anyone paid attention to Rutherford's design, with his electrons skipping lightly around the nucleus. However, two years later Bohr picked up Rutherford's ideas and enhanced it."

"This must have been an exciting time," I said.

Albert nodded. "Indeed. Scientists were feverishly trying to figure out exactly what was inside the once indivisible atom. After some confusion, they eventually sorted out their ideas into what is called the Standard Model."

"My goodness," I said, "it does sound like they all actually agreed on something. Wonders will never cease."

AT LAST, A STANDARD

The nice thing about standards is that there are so many to choose from.

—Author unknown

I'VE BEEN REALLY ENJOYING our little jaunts, Albert," I said, "but I'm afraid that they're getting kind of boring, and I'm not sure what we should do going forward."

"What do you mean?" he asked.

"There's nothing much to see any more," I answered. "It was fun watching Marie Curie working at her lab bench. Her early years, I mean—not the time when she became ill. But now, with Rutherford and those guys, there's nothing much on show."

"That, dear boy," Albert said, "is because it's all starting to be unseen. The deeper secrets of matter are no longer visible to the eye. No more uranium glowing in the dark to amuse you. It's all in the mind now."

He paused, then said, "There are always particle accelerators, like the ones at CERN in Switzerland. We could head over there in a swirl of mist, and take a look? You could get a T-shirt with a picture of the Large Hadron Collider on it?"

"No thanks," I said. "Enough with the mist. Maybe you could just cut to the chase and tell me the latest ideas on the composition of matter."

Albert shrugged. "Okay, but that's a pity. Working the mist is the most fun thing I get to do now.

"Still, let's move on. If you've been paying attention up to now, then you know that we had figured out that the atom had three very tiny things inside?"

"What is this, a quiz?" I asked. "That's easy. Protons and neutrons in the middle, with electrons whizzing around them."

"Quite so," Albert said. "Or maybe I should say not quite so. But let's leave the complications for later.

"We thought that those three were the basic building blocks for all matter, right? Until the results coming in from particle accelerators—those machines you didn't want to look at because they were so boring—turned up what looked like hundreds of different sub-atomic particles. At one point, the researcher Murray Gell-Mann wrote in his book, *The Quark and the Jaguar*, that there might be an unlimited number of such particles."

"Unlimited?" I said.

"That was just one of Murray's ideas," Al said. "He was in good company, because none other than Chairman Mao entertained the same idea."

"Mao was a scientist too?"

"Not really," Al said, smiling at me. "It's just that he saw his version of communism as being rooted in the very nature of the atom. He was so bound up in this that an American Nobel prizewinner, Sheldon Lee Glashow, proposed in 1977 that the building blocks of all matter be called *MAONS*—to honor Mao for stressing the fundamental unity in Nature."

"Typical," I mumbled. "So, we have prominent people imagining that we could have an infinite number of fundamental particles? Just as no two snowflakes are alike?"

"You really shouldn't be dabbling in physics, you know," Albert said. "It is theoretically possible for two snowflakes to be identical. It's just that they usually aren't."

"I stand corrected," I mumbled.

"It was in the 1960s," Albert said, "that physicists began to understand how the atom really worked. They realized that its protons and neutrons were made up of still smaller objects. They called them *quarks*."

"Like in the title of Murray's book?"

"Exactly," Albert replied. "Murray and Richard Feynman were thinking of a trio of particles, and Murray called these little things quarks, thinking that he'd invented a cute word. He later discovered that the word had already been used in *Finnegans Wake*, a book written by James Joyce."

"A tiresome writer, Joyce," I said. "I just don't go for that stream of consciousness crap."

"Whereas your books are so much clearer?" Albert said.

"Was that sarcasm?" I said.

"Possibly," Albert said. "So, there are six quarks, and they have ludicrous names, just to annoy people. Those are: *up* and *down*, *charm* and *strange*, *top* and *bottom*. It took until 1995 to confirm the existence of them all."

"Those names are pretty weird, aren't they?" I said.

"It could have been worse," Albert said. "For a time, the top and bottom pair were known as *truth* and *beauty*. Thankfully, that didn't catch on."

"How do they . . . work, if that isn't a stupid question?" I asked.

"Oddly, no, not so stupid," Albert said. "Ups and downs are found in protons and neutrons. The other four account for the apparently huge number of items discovered in particle accelerators. Those hits turned out to be made up of just charms, stranges, tops and bottoms.

"There is another group of subatomic particles, however, called *leptons*. Six of those too. And there is one of those that you, yes even you, know already. It's the *electron*."

"What do they all do?" I asked.

"Three of them have an electrical charge," Albert said. "The electron, the *muon* and the *tau*."

"More stupid names," I said. "Muon and tau."

"For once, I agree with you," Albert said. "Now, the other three leptons are called *neutrinos*, and they are electrically neutral."

"So," I said, "let me get this straight. The up and down quarks, with the lepton that we call the electron—they make up matter?"

"Sort of," Albert said, "I see that your brain is beginning to grasp the *Standard Model*. But the particles need something else to make it all happen. Can you guess what that is?"

"Some kind of attraction," I said, "like a strong glue to keep them together. After all, we need to use those humungous particle accelerators to smash atoms apart, right?"

"Hmm," Albert said, "I'll have to write to CERN and tell them that their accelerator is—how did you put it, humungous? They'll be so impressed with your description."

"So, there is a glue, right?" I said, trying to keep Albert on track.

"Not a glue, but a force," Albert said. "Or, more accurately, four different forces. Although, some speculate that there might even be a fifth."

"Let me guess," I said. "Gravity is one of them?"

"Yes, that's the one that most people understand," Albert said. "You know gravity, because you experience it every second of your life. It keeps you attached to our planet. Funny thing, though . . . gravity is the force we're most aware of, but it is comparatively weak. The other forces are much, much stronger.

"You also know the one called the electromagnetic force, because you've seen things like magnets, and what happens when you hold them near iron filings, right?"

"Yes," I said. "Compasses too. I remember playing with those as a boy."

"So did I," Albert said, flashing me a reluctant smile. "My father gave me a compass, and I sat with the thing for ages. I just couldn't get over how fascinating it was, and how there must be something special going on deep inside it. I learned later about electromagnetism, and why it keeps electrons whizzing around the nucleus of the atom. Consequently, why the world is made up of stuff."

"We have two forces, then," I said, "gravity and electromagnetic. But you said there were four."

"These will be harder for you to understand," Albert said, "because you don't experience them directly. First, there's the strong nuclear force, that keeps the protons and neutrons together. Then the weak force is even harder to understand, although it turns out to be required for radioactive processes to function, among other things.

"The forces differ in how far they seem to act. For example, the weak and strong forces only seem to operate inside the atom, whereas the other forces, like gravity, seem to operate at a distance."

"Okay," I said, "I get the general idea of this so-called *Standard Model*, with its quarkies, its muons and Klingons, and that's all I really need for my purposes. Perhaps we could now move on, and look at some other areas I'm interested in?"

"A general idea isn't enough," Albert said, his expression troubled. "There's more to it than that. The Higgs boson, for example, and all the antiparticles. You need to dig deeper into the Standard Model, and to understand how the particles behave: what's called their spin, their charges, and so on.

"I also think you need to study several other fields of theoretical physics, like quantum gravity, string and superstring theory, supersymmetry, and so on. Otherwise, you won't ever be able to understand any of the so-called Theories of Everything."

"I don't think I need that level of detail," I said. "In fact, I don't particularly care about the specifics of the various particles. I consider them to be . . . well, clutter."

"Clutter?" gasped a horrified Einstein.

"Yes," I replied. "I don't think it really matters what the particles are. Tell me, Albert, how many theoretical physicists at universities are working on a theory of everything?"

Al frowned. "Hardly any, I'd say. They're working on very specific issues that are parts of the puzzle."

"I'm perfectly happy for them to keep their noses to their grindstones," I said. "I just don't have the time for all that."

"But," Al objected, "you need to understand the precise and very specific nature of how particles make up what we are. If they weren't in their existing configuration, then we wouldn't even exist."

I smiled at him. "I can see it's time for Ping Pong Mountain.'

"Ping what?"

"Al, imagine a mountain with a huge pyramid of a million ping-pong balls on top. Each is individually numbered. At the foot of the mountain, a man is standing, enjoying the spring breeze."

"What does this have to do with anything?" Albert asked.

"The breeze disturbs the balls, Albert, and the whole pile starts to collapse. The ping-pong balls go tumbling down the mountainside. Some of them get tangled up in the brush, others fall into crevices here and there. As the cascade goes further down the slope, there are fewer and fewer balls continuing their run."

"What's the point of this?" Al said.

"The man at the bottom hears the noise and turns around. He sees only one ball rolling towards him, because all the others have been blocked along the way. The man bends down and picks up the ball, looks at it and says, 'My goodness, what is the very special reason that this ball, numbered 472,887, has arrived into my hand?'"

"There's no special reason," Al said. "That just happens to be the ball that made it all the way. It could just as easily have been any other."

"Exactly," I said. "When people tell us that the specific particles in the Standard Model are just magically the right mix to produce galaxies, planets and people, and that this couldn't have happened by accident, I have to say that it could just as easily have gone another way."

Al nodded. "In which case, we probably wouldn't be here to talk about how special it all seems."

"Right."

Al chuckled. "Ping Pong Mountain gives you a little problem, Ian. Who built the pile of balls in the first place? Who was the first cause, or creator of that situation?"

"I'm working on that, Al."

"So, what do you want to learn about now?" he asked.

"I want us to move from the very small to the very large," I said. "I want to look at how the universe came to be."

"Okay," Albert said, "I can help you with that, I think." He paused, then looked at me rather sheepishly. "Before we move on, however, there's something I should have told you about sooner."

"What's that?"

"The dreaded quantum," Albert said.

THE EERIE QUANTUM

The atoms or elementary particles themselves are not real. They form a world of potentialities or possibilities rather than one of things or facts.

—Werner Heisenberg

ALBERT LOOKED ACROSS AT ME. "That's right. Before we look at the cosmos, we should talk about what they call quantum theory. I'm afraid I've been holding it back from you. I'm not a fan, you see."

"I don't understand," I said. "It's science, just like everything else we've looked at, isn't it?"

"Not in fact," Albert said. "I became known for saying that it's loopy. I still think that, by the way. It's an abomination, and I find it hard to accept that the universe follows quantum behavior."

"I read about that," I said. "You fought the quantumistas for years, didn't you?"

"Decades, actually," sighed Albert. "It was a constant source of misery for me that I couldn't disprove their ideas."

"I've read about quantum theory in a few books," I said, "but I've never quite understood it."

"Nobody does," Albert said, "and quite a few have gone mad thinking about its implications. There's that famous saying that explaining quantum theory to someone is like teaching an Eskimo to speak French, but without actually speaking French."

"Try to explain it to me, then, *s'il vous plaît*," I urged. "After all, you've taken me this far. You need to show me everything, don't you?"

"Okay," Albert said, "but this is against my better judgment. There again, driving you crazy now will just be saving time. You're already well on the way.

"There are a lot of players in the quantum story. First, you might know that it was Newton who, in 1666, showed that light could be split into seven colored bands. You know those, right?"

"I think so," I said. "Red, orange, yellow, green, blue, indigo, and violet."

"Exactly," Al said. "Newton settled on there being seven colors because of his fascination with the seven notes in the western musical scale, and the seven days of the week."

"I didn't know that," I said.

"Of course you didn't. Now, it took until 1800 before another key realization was made, by William Herschel. He was a violinist, just like me, and he also composed music."

"What's his scientific contribution?" I asked.

"He was also using a prism to split light and, when he moved a thermometer through the colors, he found that the temperature of each band was slightly different. He also noticed that, just past the red end of the range, it was different again. He had discovered what became known as *infrared* radiation—a type that humans can't see. A year later, Johann Ritter used a different technique to find out that this was also true at the other end of the colors, beyond violet. That was, of course, *ultraviolet*."

"How is all this connected to quantum theory?" I asked.

"This idea of light colors being of a different temperature wasn't new," Al said. "It had long been understood by those in the pottery business that it didn't matter what material you used to build your kiln. They knew they had to watch while their ovens got hotter and hotter, first emitting a cherry red, then a blue-white color. It was the color that told them when the temperature was right to begin work.

"A researcher called Gustav Kirchhoff started to think about measuring the levels of this radiation. He theorized about something he called a blackbody—an entity that would absorb all levels of radiation. At low temperatures, it would emit nothing and would appear black, but when hotter would show visible emissions."

"I'm still not getting the quantum thing," I said.

"In the 1880s," Al said, ignoring my objection, "German companies were keen to find a way to measure the intensity of radiation, the reason being that they wanted to invent electric light bulbs better than the competition's. The problem was that nobody had a way of measuring the intensity of light.

"In 1983, a young man named Wilhelm Wien found a formula that seemed to work, at least for shorter wavelengths. He would go on to win a Nobel prize in 1911. But as early as 1899, it was noticed that the formula wasn't perfect. Enter Max Planck."

"Aha," I said, "I've heard that name."

"Max tinkered with various formulae and ended up with something that appeared to fit. He was frustrated, though, because he didn't understand exactly why it worked. He was also very disturbed by the implications of his discovery."

"Still," I said, "the electric companies got the tool they needed to measure the output of their light bulbs."

Al nodded. "Planck worked on, and eventually made an announcement on December the 14th, 1900—that's considered to be the birth of quantum mechanics. He said that energy buzzing around couldn't do so smoothly, but could only be issued in packets of definite size. He called those *quanta*."

"I see," I said. "That must have caused quite a stir."

"Not at all," Al said, smiling at me. "Hardly anyone paid the slightest attention, and it wasn't considered at all significant. Most thought that classical Newtonian mechanics still applied, and that all we had to do was somehow fold in that quantum idea and move on."

"What changed that?" I asked.

"I had a role to play there," Al said, chuckling. "In 1905, I stated that light itself was quantized. It too could only be issued in discrete packets."

"What happened then?"

"It was Bohr who was the next to rattle our cages. He applied the quantum to the workings of the hydrogen atom, to explain the glowing of colors in a discharge tube. That finally did get everyone interested. Perhaps, they now thought, this quantum idea was something quite fundamental in the overall scheme of things.

"After the First World War ended, people really started to work on these ideas. In the 1920s, a new theory of wave mechanics was developed, pulling together the work of a bunch of different people—Heisenberg, Born, Jordan, de Broglie, Schrödinger, and Eckert."

"The quantum gang was taking over," I said.

"Correct. It was much to my dismay, and I never quite accepted it," Al said, frowning. "I recall writing to Born, saying that something didn't feel real about it."

"Yet it was?"

"I'm afraid so," Al said. "People were using the concepts to solve all sorts of physics problems. The new theory of quantum mechanics was holding up very well indeed."

"You were disappointed," I said, looking at the old man. "You wanted a mechanical universe."

"I didn't get it," Al said. "As it happens, I've been keeping an eye on this from on high. It turns out that the fundamental building blocks of the cosmos are not just particles, but are more fluid. In fact, they also act like fields that ripple throughout the entire universe. A field will have a value at every position, and those values change in unpredictable ways."

"But you said that light was emitted in packets," I said. "How does that square with the cosmos being made up of fluid fields?"

"That's what they now call quantum field theory," Al said. "If you look at, say, the single electron field that emanates through the universe, you'll see that it's made up of tiny packets. The electrons, in fact. There are also similar cosmic fields for the quarks, and so on."

"How many fields are there?" I asked.

"A few," he replied. He sighed. "At least, that's what they're saying right now. Most think there's a field for each of the fundamental particles in the Standard Model."

I smiled. "All of the fields rippling and pulsating together, without any seeming order or discipline. Disgraceful."

"You're poking fun at my Newtonian beliefs," Al said, frowning. "You should know, however, that there are even those who suspect there might only be one field, not several. They think that the fields only appear to be different because of the way we're looking at them. Those folks point out that the mathematics for each of the supposed fields is almost identical."

"You're right, Al. This is eerie stuff. It would mean that everything in the universe is interconnected in one giant wobbly energy field."

Al grunted. "You can see now why I'm so upset. This is no way to run a cosmos."

THE BIGGER MATTER OF THE COSMOS

In which we turn our gaze to the large.

WHAT'S IN THE SKY?

Of all things visible, the highest is the heaven of the fixed stars.

—Nicolaus Copernicus

I WAS HOVERING BESIDE ALBERT AGAIN, but I didn't recognize where we were this time.

"This isn't Paris, that's for sure," I said.

Albert chuckled. "You never know, it could be," he said. "We're in the year twenty thousand BCE."

"Why this era?" I asked. "What could we possibly learn here about the cosmos?"

Albert laughed. "We can gain assurance by seeing how we humans have always had one thing working in our favor, namely that we are curious. That's not just a select few of us either; most of us have always been curious about the world and its contents. I wanted you to see what our ancestors were up to, tens of thousands of years ago."

"It doesn't look so bad down below," I said, looking at the idyllic scene. "Mostly unspoiled grassland. And I can see some caves, on that ledge there, about fifty feet above the valley floor."

Albert nodded. "We're about to see Herman, the head of a family living in one of the caves. He's been woken up by the early daylight filtering in from the outside world. About now, he's throwing off the bearskins and getting up from his bed, trying not to disturb Wendy, his domestic partner. Look, here he is, coming out onto the ledge."

"Who's the guy exiting the cave next door?"

"That's Cyril, his neighbor," Albert said. "Let's listen."

"Morning, Cyril," Herman said.

"Back at ya," Cyril replied. "Looks like it might be a nice day for hunting, don't you think?"

"You're right," Herman said. "One of the guys said the antelope are back in the valley, maybe just a few miles away. We should get a group together and go get some meat."

The two men were quiet for a while, with both looking around at the valley, and up at the sky.

"Have you ever wondered," Herman said, "what the sun is?"

"Can't say I have," Cyril replied. "Have you?"

"A bit," said Herman. "I mean, the sun comes up every day like clockwork, doesn't it?"

"What's that mean . . . clockwork?" asked Cyril.

"I dunno," Herman said, with a shrug. "The word just came to me." He paused, then said, "So, I've been wondering what the sun really is. Is it a god of some kind, flying across the sky, or what? I've also been thinking about the world itself. Everybody used to say it was flat, and I must say that's what it looks like to me. But now there's them that says it's round."

"Can't be round," said Cyril, gazing confidently out at the valley. "I mean, just look at it. It's flat as far as you can see."

Herman nodded. "I suppose. But if it's flat, then how come we can't see further than we do? Shouldn't we be able to see all the way to the edge?"

"Well, that's our eyes," Cyril said. "Our eyes aren't that good, are they? They can only see so far. You know how our dogs see things much further off than we can, and eagles and hawks seem to pick out the tiniest mouse from way up in the sky? Our vision's just not good enough, that's all.

Anyway, Herman, what's all this fuss about the sun? Has it got anything to do with that new priest of ours?"

"He makes me think," Herman said. "I've been watching what he's been doing. He got some of the boys to make a clearing, and he's placing stones all around: in the center, and in a semi-circle. Says he's going to keep track of the dawn light and note where it shines on the stones. Thinks he can learn about the sun, and maybe even predict the changes of the seasons. Might even be able to explain why the stars move."

Cyril frowned. "He'd better be careful, then. Some of us don't think it's a good idea to look too deeply into the heavens. Maybe the gods will get angry."

"Not exactly brilliant examples of human curiosity, are they?" I said to Albert. "They don't seem to be too motivated to figure out the actual answers to their questions. Sounds like they're more interested in getting their hunting party together, grabbing their spears, and killing things."

"Good point," Albert said. "I must say, however, that I quite like these two men. I've watched them before you know. They're excellent hunters, bringing home the meat for their families, and sharing it with some of the older people in the tribe."

"But they didn't know much else, did they?" I said. "That business about the world being flat?"

"They were in good company," Albert said. "It took a long time for the consensus to build that the world was round. We're smarter now, aren't we, and our settled science holds that our planet is indeed round." Albert smiled, "At least, for now."

He continued. "Our ancestors also had another embarrassing belief—that the sun moved around their world."

I chuckled. "Well, that's what it looked like, didn't it?"

"True," Albert said, "but it has always puzzled me why it took so long for anyone to wonder what it would look like if the earth went around the sun."

I thought for a moment. "That looks exactly the same," I said.

"Precisely," Albert said. "But let's look at the sun later." He paused, then said, "You know, Herman and Cyril's little priest had quite an enterprising mind. He kept careful records of what he saw in the skies, making notches in sticks and chiseling marks into stone.

"And even if he didn't figure out how the stars seemed to move around, there were others who did. I think we should visit ancient Egypt."

CONSTELLATION REALIZATIONS

Consider the beauty of life. Watch the stars and see yourself running with them.

—Marcus Aurelius

ALBERT SQUINTED AT ME. "Those cavemen we just saw probably noticed that the twinkling lights in the night sky seemed to shift around as the days went by, and that there seemed to be patterns made by some of the brighter stars. Our two friends might have been intrigued to learn that this was something that the Egyptians were investigating closely. And very early indeed in human development."

"How early?" I asked.

"Depends on the researcher," Albert said, "some say two or three thousand years BCE, while others think ten or more."

"It's just occurred to me," I said. "You could settle that, couldn't you? You could do your mist-swirling thing and go back to find out the truth. Then you could correct all the errors in our history books."

"That's true," Albert said, "but I am not allowed to do that. It's one of the conditions of being a Time Lord. Moreover, giving people the truth doesn't necessarily change their minds. What is that old saying? *Don't confuse me with facts?"*

"I see," I mumbled. "So, what are you going to show me now?"

Mist swirled, and when it cleared, we were looking down on a small group of what looked like ancient Egyptian priests. They were gazing up at the night sky, tilting their heads back to sight through instruments of some kind.

"What is your assessment, o wise one?" said one of the younger priests standing off to the side. "Has the transition occurred?"

"What transition?" I said, whispering to Albert.

Albert nodded. "You should know that the Egyptians made some truly astounding discoveries by studying their night skies. Every evening, those old-timers would have looked up and seen one of the twelve constellations of the Zodiac.

"You know those, right? We in the west call them *Aries*, *Taurus*, *Gemini*, *Cancer*, and so on. Those star systems would gradually shift across the sky, succeeding each other as the months went by, and a very interesting thing happened at the *vernal* equinox. That's what we in the northern hemisphere think of as the spring equinox."

"What's an equinox?" I asked.

"God help us," Albert said, in disgust. "An equinox—which you should know if you paid attention in high school—is one of four key days in the year: the two equinoxes where the amount of daylight equals the darkness, and the two solstices, which are the in-between dates.

"During the spring equinox, the Egyptians would look up and see a particular constellation in place. But they also noticed that, over extended periods of time, the vernal equinox constellation would gradually shift, and be succeeded by the following one in the series.

We're talking very long periods here, since it takes about 2,160 years for one vernal equinox constellation to shift to the next."

"The change is not yet complete, my son," said the elderly priest, as he lowered his sighting instrument. "I know you're keen to declare the new era. But it is difficult to be precise as to the changeover."

"Why, master?"

"Because while the stars themselves are bright, the constellations that hold them are mostly dark, and one cannot tell where one constellation ends and another begins. That is why the pharaoh trusts me to determine these important changes with my calculations. And what I can tell you is that the new age will be upon us in about three years."

"Oh," said the young priest, clearly disappointed.

The older priest smiled. "Cheer up, my son. Even though the transition is not yet here, we must still go and alert the Pharaoh at once."

"Why the urgency, master?"

"Because our stone masons must begin creating the symbols for the new era. It will take them several years to carve all the monuments for the age of the lion."

"This transitioning effect," Albert said, "is known as the *precession of the equinoxes*. And, although it doesn't really impact our daily lives much, the Egyptians thought it very important indeed. They celebrated this by making statues and inscriptions that used the images and themes of the vernal constellation of their time."

"Makes me think," I said. "Christians used one of those Zodiac signs, didn't they?" I asked.

"Yes," Albert said. "They often say that the fish was adopted as their sign because Jesus was a fisher of men. But it could just be that ancient Christians chose the fish as their symbol because it was the vernal equinox constellation of their time, and they thought the stars were announcing the arrival of their messiah."

"What's the big deal with a change of constellation, anyway?" I asked.

Albert smiled. "Some say that the beginning of a new vernal constellation will bring about major changes, like the fall of civilizations or empires."

"What era are we in now?" I asked.

"Some say that we are close to the Age of Aquarius," replied Albert.

"Wait," I said, "I've heard of that. It was back when I was living in London, in the 1970s. A musical called *Hair* was all the rage, having arrived in England after a run on Broadway. Say, Albert, could you do me a favor, and swirl your mist thingy to take me back to that time in London. I'll show you what I mean about Aquarius."

"This is highly irregular," Albert said, "but, as a treat for you, I'll do a rewind, and you stop me when it's the right moment in your life."

The mist swirled, and I saw my life winding back in time.

"There," I said, "there I am, coming out of the office in Oxford Street, just after work. My goodness, Albert, I still have that gold-capped umbrella that I'm swinging as I walk along. And look, I'm about to meet my friend, Alice, for a drink."

"We just have to see this musical, Ian," Alice said. "Everyone's going. The show has an important anti-Vietnam War message."

"Sorry," I said, "I'm just not that interested in hippie philosophy. Maybe you should go with your other friends?"

"But you'll be missing something," Alice said. "Did you know that some of the cast appear naked at the end of the show?"

"Who exactly?" I said, my interest growing for some reason.

She nodded. "Just to be clear, they don't all appear naked. Just those who feel the spirit moving them on the night. They see it as their way of communing with the universe." She paused, then added, "Just think, you might see some pretty young things . . . starkers, right there in front of you."

"Hmm," I said, "well I must say it has been a while since I went to a good musical. Okay then, in the interests of art and philosophy, I'll get us tickets."

"One evening shortly afterwards," I said, turning to Albert, "we watched the show in the Shaftesbury Theatre. Luckily, the cast's cosmic spirit was alive and well, and several of the young things got their kit off."

"Got their kit off?" Albert said.

"A British colloquialism," I said, with a smile. "Means they got naked. I was almost bowled out of my seat by dirty old men behind me, lunging forward to get a better view of perky young mammories. The only reason I didn't lean forward myself was that I was a dirty *young* man, with effective eyesight."

"I see," Albert said. "And did you enjoy the show? Apart from the breasts, I mean."

"Oh yes," I said. "I recall that one of the evening's melodies was the song *Aquarius*, at the opening and then reprised near the end. Quite a hummable ditty and, in the days that followed, I went around London singing it to myself, blissfully unaware that I had just been manipulated by left-wing propaganda.

"And you know what you said, Al, about a new age causing disruption? Might be true, because the show's run in London came to an abrupt end when the roof of the Shaftesbury Theatre caved in."

"Naughty Aquarius," Albert said, chuckling. "So, yes, our Egyptian friends were darned clever, what with being able to figure out how precession worked. Their work is an example of just what can be done if people make careful records over time, and then look for patterns.

"But the Egyptians weren't the only brainy ones around. Let me tell you about one very, very intelligent man—a Greek, as it happens.

His accomplishment is an example of what can be done with a flash of brilliance. He did something that will take your breath away."

HOW BIG?

Talent hits a target no one else can hit.
Genius hits a target no one else can see.

—Arthur Schopenhauer

WHO IS THIS SUPPOSED BRAINIAC you wanted me to see?"

"A certain Eratosthenes," Albert said. "He did his thing in the third century BCE, was a very successful man and, at a certain point, crossed the Mediterranean to go from his home in Greece to Alexandria. He went to take up his new post as that city's head librarian.

"You might think he'd feel homesick in North Africa, but Alexandria at the time was a thoroughly Greek city, with Greek architecture, language, clothing, and so on. He'd have felt right at home there. Let's watch him."

The mist swirled, then I was looking down at a distinguished looking gentleman, seated at a table in a room where the shelves were full of scrolls.

"This is the famous library at Alexandria?" I asked Albert. "The one that the Romans and the Muslims eventually destroyed?"

"That's right," Albert said.

After a minute or two, I whispered, "This . . . Eratosthenes, he doesn't seem to be doing anything."

Albert nodded. "That's because he's a great thinker." There was a pause, then Albert said, "Let me ask you this—suppose you went in there and Eratosthenes interviewed you for a job in his library, how do you think you'd fare?"

I thought for a moment. "Well, I'm a twenty-first century man. I have two millennia's worth of knowledge that he lacks. Shouldn't it be the other way around, with me interviewing him?"

"Hmm," Albert said. "Suppose, then, that the test he gave you was to devise a method for finding prime numbers. What would your answer be?"

I looked up at the sky. "Well, I'd tell him that I'd need some form of computing power. These days, I'd use massive banks of computers, and devise searching algorithms to find and check for the primes.

"In his day, there were no computers, so I'd probably ask him to assign me dozens of support staff, equipped with abacuses." I paused. "Or is the plural *abaci*?" I smiled. "You know what I mean, Al? Those frames with wires, and beads you can slide up and down to count with?"

Albert scoffed. "That's not what the Roman and Greeks used. They had pebbles they called *calculi* that they slid in grooves on a table. Look, should you really be delving into history? You know so little about it."

"Point taken, Albert. Getting back to Eratosthenes, I still think I should be the one interviewing him for a job."

Albert smiled. "I think that Eratosthenes, after listening patiently to your ramblings, might have said, 'Thank you, Ian, for your interest in our organization. We wish you well in your endeavors and, on your way out, please ask the next candidate to step in.' "

I was shocked and a little annoyed. "You mean I wouldn't get the job?"

"Of course not," Albert said. "You see, Eratosthenes was able to find primes using just a pointed stick and a stretch of smooth sand."

"That can't be right," I said, confused. "Finding prime numbers is difficult."

"I don't have a stick or sand," Albert said, "but here, take a notepad and pencil, and do what I describe. First, write out a range of numbers. Doesn't matter how far you go, but let's say from one to a hundred."

I did that, carefully writing down the numbers in ten rows of ten.

"Now," Albert said, "cross out the number one. It's not a prime, it's a special case. Then, circle the number two. It must be prime since there is no regular number smaller than it."

I did as he said. "Now what?"

Albert smiled. "Now cross off from your list every second number that follows two. That's because two must divide exactly into them, and therefore they can't be primes."

I put my pencil close to the page. "You mean, cross off four, six, eight, and so on, all the way up to and including one hundred?"

Albert nodded, and so I did what he asked.

"Now," he said, "go back to the beginning of your list, and find the first number that you haven't circled or crossed off."

"That would be . . . three," I said.

"Circle it," Albert said. "It must be prime. Now cross off every third number after three, since they can't be primes. If they're already crossed off, just ignore them."

I did that too, and then Albert said, "Okay, same process. Back to the beginning of your list and look for the first number not circled or crossed off."

I looked at the page. "I already crossed off four when I worked on the twos, so the next would be five," I said. "I suppose that I should circle it because it's prime, then cross off every fifth number from there?"

Albert nodded. "By the time you've dealt with your entire list in that way, all the circled numbers that remain are the primes."

"My God," I said, "that is fiendishly clever. Simple, too."

"You don't need a computer," Albert said, with a chuckle. "You just need to be able to count along a row of numbers. The system will work for any range, no matter how high you want to go. This arithmetical method is known as the *Sieve of Eratosthenes*."

"Thanks for showing me this," I said. "I do agree with you. Eratosthenes was definitely a genius, and I would be the one working for him, not the other way around."

I looked at Albert. "Where do we go to next? With the mist?"

"We're not done here," Albert said. "Finding primes was just a little party trick for Eratosthenes. He's about to do something even more impressive."

"That's hard to believe," I said. "Show me."

Albert pointed down to where Eratosthenes was, still sitting in his library room. "You know, this library he's sitting in was the center of knowledge for the entire civilized world. Alexandria was where Aristarchus declared that the earth orbited the sun. It took us well over a thousand years to agree that he was right.

"Archimedes invented the screw here—the one that bears his name, and people still use it to raise water from lower to higher levels. Euclid wrote his work *Elements* in Alexandria. Punctuation had its beginnings here, as did the idea of putting accents over letters to change the way they sound."

"It must have been quite a place," I said, now in admiration of these old-timers and their skills.

Albert nodded. "They were determined to keep their library as the world's knowledge leader. One practice they followed was to meet all the ships that pulled in at the docks. Inspectors from the library would check the holds' contents and, if there were any important writings or scriptures, would take them away to have them copied.

"This kind of activity would have kept the library staff fully occupied. But, they occasionally had time for other projects too. Let's watch as one of the staff reports in."

"Are the boys busy today, Poulos?" said Eratosthenes, looking at the assistant who'd just stepped into the room.

"Not too much," Poulos replied. "Only one vessel arrived this morning, and it had no scrolls of interest."

"I see," Eratosthenes said. "Then maybe we can take some time to work on a new project."

"Good. Is this about your method for finding prime numbers? I thought you'd completed that one. Awfully clever, I must say."

"No, dear Poulos, this is different. I am going to figure out the size of our planet."

Poulos frowned. "I heard you talking about that at lunch the other day. But how on earth could we do that? Are you going to send a team all the way round the globe, to measure the distance?"

"No," Eratosthenes replied. "If you think about it, they'd hit mountains and oceans, and would never manage to keep to a straight line. No, brute force is not my style. I'd rather use the little grey cells of my brain.

"You see, I heard that there's a place south of here called Syene. The people there say that, at the summer solstice and only on that one day of the year, the sun lights up the bottom of a deep well there. That must mean that the sun is then directly overhead."

"Okay," Poulos said, "but what of it? How does that help us?"

"Well, dear Poulos," said Eratosthenes, with a smile, "I was strolling outside here on that very date last year, and I noticed something—the pillars in our library's grounds all cast a very slight shadow. You see, we in Alexandria are further around the curvature of the earth. While the sun was directly overhead at Syene, it was not overhead for us, hence the little shadows here."

Poulos frowned. "Are you sure it was the same time of day? Or were our pillars maybe leaning a little bit to one side?"

"I considered those ideas," Eratosthenes said, "but I am sure about the time, and I had the pillars checked. They're perfectly vertical."

"I still don't get it," said a puzzled Poulos. "What does this mean?"

Eratosthenes smiled. "The solstice is coming up. I need to get an accurate reading of the shadow's angle here in Alexandria, and to know the exact distance to Syene. Then, with a little bit of geometry and some time on my calculations, I will be able to tell you the circumference of our dear planet."

Albert smiled at me. "Eratosthenes arranged for the experiment to be done. And you can imagine what he might have put together as equipment: perhaps a tripod, with a plumb line or stick hanging vertically from a central point. Or maybe a sundial of some kind. The same equipment would have been set up at both locations.

"All that remained was for accurate measurements of shadows to be taken on the same date and time of day. By doing this, Eratosthenes confirmed the lack of a shadow at Syene—we call it *Aswan* these days, by the way—and he captured the exact length of the slight shadow at Alexandria. Of course, he needed the exact distance between the two locations. Hmm, now how could he do that back in those early times? Doesn't seem possible, does it?"

"Ah, just wait," I said. "You tricked me with that prime number thing. I guess you're going to tell me that they did something clever to figure out the distance between cities?"

"If that distance problem hadn't already been solved," Albert said, "then I wouldn't mind betting that Eratosthenes would have figured out some cute and simple way.

"But early armies, who were most interested in exact distances, had already found a solution. They used a special type of cart, fitted with what's called an *Odometer of Vitruvius*. It had a hopper full of little round stones and, below that, a collection bin that began the journey empty.

"With each turn of the cart's wheel, a lever was tripped, and a single pebble was dropped into that lower bin. At the end of the journey, all they had to do was to multiply the circumference of the cart's wheel by the number of pebbles that had dropped."

"Brilliant," I mumbled.

Albert smiled. "Eratosthenes now had all his desired data, so he sat at his abacus table, worked some magic with geometry and angles, and figured that our planet is a quarter of a million *stadions* around."

I frowned. "They had different units of measure back then?"

"Yes," Albert said. "Unfortunately, we don't seem to have a conclusive definition for *stades*-that's what we call that unit in English—but the researcher Lev Vasilevich Firsov, averaging out some ancient accounts, thinks that it might have been 157.7 meters. If so, then I make Eratosthenes's circumference to be just over 24,300 miles—not too far out from the currently accepted value of around 25,000 miles."

"Well done, Eratosthenes," I said. "Color me impressed."

HELIOS, OR GEOS?

It alone remains immoveable, whilst all things revolve round it.

—Pliny the Elder

IT WAS MY TURN TO GAZE AT ALBERT, who seemed to be lost in thought.

"What's next, then?" I asked. "Where do we mist off to now?"

"Shouldn't you say, '*When* do we mist off to, instead of where?" Albert said. "This is time travel after all."

"What's the next topic then?" I said, anxious to make progress.

"You'll have to forgive me," Albert said, "I need to talk about several important people, but we don't have time to visit them individually.

"It's true, however, that they all concerned themselves with a major issue that preoccupied mankind for the longest time, namely—does the sun orbit the earth, or is it the other way around?"

I frowned. "What do we have in our records about that?"

Albert smiled. "You've got to admit that, for a lot of early humans, it looked like the sun went around the earth. It seemed to rise in the east, didn't it, then move to the west? Who knows who was the first to bring up the idea that it was precisely the other way?

"What we do know, however, is that Aristotle, who lived from 384 to 322 BCE, was firmly convinced that the sun moved round the earth.

Nonsense, thought Aristarchus of Samos, who lived from 310 to 230 BCE. You remember, I told you about him being at Alexandria?"

"Yes, of course," I replied.

"He was just as sure that the earth moved round the sun."

"Well done, Aristarchus," I said, with a smile.

"Sadly, he was overruled," Albert said. "Claudius Ptolemaeus, in the second century CE, made such a fuss about this that we still talk of the geocentric model as being the *Ptolemaic System.*"

"So, they were back with the earth being the center," I said. "Their heads must have been spinning by now. What to think?"

"As sometimes happens," Albert said, "it took a while, but eventually it was Nicolaus Copernicus who came out and pushed for a heliocentric model. He published his ideas in a work he titled *De Revolutionibus Orbium Coelestium*, published in 1543.

"It wasn't just that he declared the sun to be the center of our galaxy. He also disclosed his computations, which turned out to be very effective at predicting the future positions of planets. Folks found that to be quite convincing, and so listened attentively to his ideas."

"The issue was settled then," I said.

Albert smiled. "There was a teensy problem. You see, the Church knew that the Bible stated, without doubt, that the earth did not move. Not round the sun, not round anything. Its position was fixed, and there are several verses to support that. In First Chronicles, 16:30, for example: *Fear before him, all the earth: the world also shall be stable, that it be not moved."*

I laughed. "Well, if I'd been Pope back then, the solution would have been simple. Just get hold of that heretic Copernicus, and throw his ass in jail. Better yet, burn him to get rid of his evil."

Albert nodded. "You're right, but the Catholic Church somehow just ignored Copernicus and his ideas. Then Galileo came along and Pope Urban VIII, perhaps expecting a favorable outcome, asked

Galileo to publish his thinking on the arguments for and against heliocentrism. Sadly for the Pope, Galileo's writing seemed to favor the sun as the center of our galaxy. He was tried for his heresy in 1633, and was subject to house arrest until his eventual death.

"Bit by bit, however, the world accepted the idea of a sun at the center of things, and the weight of that collective opinion caused the Church to quietly drop its objections to the theory. Of course, the Church being sneaky clever, claimed that it was God himself who chose to reveal the truth through His servant Galileo."

"What about how they persecuted the poor man?" I asked. "How did they excuse that?"

"They didn't," Albert said. "They just don't speak of that any more. Don't you know anything about how the Church works its devious ways?"

I frowned. "Galileo finally did the trick then. He put the nail in the coffin of geocentrism?"

"He did," Albert said. "Now if you don't mind, we can twirl a little more mist. I want you to see someone whose thinking is very instructive in areas like this."

When the light cleared, I was hovering beside what looked like an abbey.

"This is fourteenth century England," Albert said. "And we're at a Franciscan abbey. That man down below is named William."

"Is he famous somehow?" I asked.

"It might mean more to you if I called him Friar William of Ockham."

"Oh," I said, smiling, "Occam's Razor—his technique for solving problems. I'm very pleased to see this man. You know, Albert, in my very first book—not a very good one, I admit—I tried to apply his technique to the life of Jesus. You see, the way Occam's Razor works is to cut away all complexity, and—"

"I know perfectly well how it's used," Albert said. "I just want to point out a couple of things. First, William never called his method *The Razor*. That was done much later. And what he proposed wasn't even new, because there had been similar statements made in previous centuries."

"What's your point with this story?" I asked.

"My point is that Occam's Razor is not foolproof," Albert said. "He is basically telling you that when faced with a puzzle, you should favor the solution that makes use of the fewest possible complications."

"I know," I said. "In my *Curious Evolution* book, I argued that the most likely explanation of a man dying on the cross and coming back to life is . . . that he didn't die."

"I happen to agree with you there," Albert said, "but look what happens when you use the Razor on the issue of heliocentrism and geocentrism. If you're an ancient human, looking up at the sky, the most obvious explanation—the one with the least complications—is that the sun is spinning round the earth. You seem to see the sun moving, and quite clearly too."

"Point taken," I said. "Occam's Razor isn't foolproof."

"And for one of the biggest issues of the last couple of thousand years, it gave us the wrong answer. The only reason I point this out is that one has to be careful, that's all."

"What?" I said, "you mean, when I start to develop my conclusions about how the cosmos came to be without a creator?"

"I mean exactly that," Albert said. "Sometimes, what you think is obvious, is wrong."

"Well, then," I said, "consider me duly advised."

I continued to look down on Occam for a time, happy to have seen this worthy man at work.

NOT JUST A COOKIE

I defy gravity.

—Marilyn Monroe

WHERE TO BEGIN WITH ISAAC NEWTON? Well, he was a fine Englishman who lived around the end of the 17th century, and he's credited with quite a few breakthroughs in the field of physics.

They say that, even now, space probes can be guided perfectly to the moon and other distant bodies, just by using the laws that Newton discovered.

"Of course, there's that gravity thing," Albert said, hovering beside me. "That was one of his greatest hits. And look, there's the great man himself, sitting under that apple tree."

"Who's that beside him?" I asked, looking down at the scene below.

"Not sure," Albert said. "Maybe we'll find out."

"Oh, that hurt," said Isaac.

"What happened?" asked the man sitting beside him.

"Didn't you see, Nicolas?" Isaac said. "An apple fell on my head. I knew it was a mistake to leave the house. Not to mention that we might catch the plague out here."

"Serves you right, getting bonked on the head," said Nicolas. "I wondered why on earth my clever friend picked an apple tree to sit under. Didn't you learn anything when you were a farmer years ago?"

"I wonder why it fell," Isaac mumbled.

"Apples tend to do that, you know," said an unsympathetic Nicolas.

"No," Isaac said, "I meant . . . what agency or force caused the apple to fall?"

"It's the stem, I think," Nicolas said. "As the apple ripens, the weight grows heavier. And the stem holding the apple to the tree reaches a point where it can't cope. Ergo, a falling apple."

"We're not on the same page, Nicolas," said Isaac. "What I'm really asking is . . . why does anything fall at all? What makes things drop to the ground?"

"It's in the nature of things to do that," Nicolas said. "You'd have to ask God."

"Not helpful, dear Nicolas," said Isaac. "Perhaps you could wander back to the house and check that they're getting dinner ready for us all? I'd like to think quietly about this issue. Maybe, just maybe, I can figure out why the apple fell as it did."

"Suit yourself," said Nicolas, rising to his feet. "But you might just want to move over to the shade of that oak. Falling acorns are less of a problem, wouldn't you say?"

Nicolas turned and walked towards the house, leaving his friend mumbling to himself.

Albert smiled. "Isaac eventually concluded that there was a force operating that caused things to be pulled downwards. Yes, he reasoned, every physical entity has the ability to attract other entities. So when a stem weakens and breaks, the earth below the apple tree is exerting a pulling force on smaller objects. That's why the apple falls. It's why anything falls to earth."

"Not sure I agree with Isaac about that," I said.

Albert laughed. "As if you could second guess Newton. You know, he managed to make his calculations about mass and motion work in most useful ways. He made many advances in mathematics and was among those who developed calculus. He also studied optics and the refraction of light. He eventually wrote down some of his ideas in a book called *Philosophiæ Naturalis Principia Mathematica*.

"Maybe I should give that a read," I said.

"It's in Latin," Albert said.

"Not to worry," I said, "I studied Latin for six years."

"Oh, really?" Albert said. "Then what does this mean: *Honi soit qui mal y pense?*"

"Trick question," I said. "That's not Latin, it's French, and it means something like . . . shame on anyone who thinks evil."

"Not bad," Albert said, grudgingly.

"I studied French too," I said.

"So," Albert said, "perhaps not a complete moron after all."

"Praise indeed," I mumbled.

"Let's recap," Albert said, "what did you learn in school today?"

"You're treating me like a child?" I said. "Okay, I'll go along. Today I learned that people are smart. Doesn't matter whether they're living now, or hundreds or thousands of years ago. And it doesn't matter what tools they have at their disposal. They can be darned clever. Thank you for showing me that."

"You're welcome," Albert said, "and that's a very important thing to keep in mind. As we move on, you're going to see more evidence of just how smart people were."

"What's next?" I asked.

"I think we should look at some stars," Albert said.

"What, like in a movie?" I said.

"Oh dear," Albert said, with a sigh.

HE HAD A LITTLE LIST

As some day it may happen that a victim must be found, I've got a little list.

—Gilbert and Sullivan

ALBERT WAS VERY QUIET. I watched him for a few seconds, then just had to ask, "Is there a problem?"

"I'm just trying to figure out what to show you next," he replied. "We've looked at early humans and their fascination with the skies and the stars. Then we looked at the big issue of heliocentrism, and then gravity.

"We can't go and visit every scientist who added to our knowledge of the cosmos, so I'm trying to pick out some of the more interesting characters as part of your education."

He smiled. "I've got it," he said, and then he twiddled his fingers, and we went into the mist.

"Paris again?" I said, as we emerged into a clear night sky.

"Yes, in the eighteenth century this time," Albert said. "This is the Hotel de Cluny—it's now a museum—and the man at the telescope is Charles Messier, who's doing an interesting thing. He's making a kind of catalog of things in the night sky."

"Why?" I asked.

"We think he started to make his list of objects so that he wouldn't be distracted by them as he studied comets in the night skies."

"I see," I said. "How many things did he find?"

"Let's listen in," Albert said. "He's working with his assistant, Pierre Méchain."

"You found another one, Pierre?"

"I did, Charles," came the reply. "How many does that make?"

"This new one would be M45 on our list," said Charles. "You know, Pierre, this kind of work could probably go on for some time. But I think we should publish our list as it stands, just to help others who are doing similar work."

"I agree," said Pierre. "This version can be the first edition of our list. As we find more, we can always republish."

"He was right," Albert said, "others had been doing similar work. In 1654, Giovanni Battista Hodierna had put out a list of about forty celestial objects, but it's doubtful that Messier even knew about it. Ancient Chinese astronomers also had their own lists."

"Charles," said Pierre, "what do you think these objects are?"

"Nebulae, some of them," replied Charles, "probably just clouds of dust. As for the others, I'm not so sure. I just want to be able to know where they are, so that they don't interfere with our other research."

"Do you think we'll eventually identify all of the objects that are up there?" said Pierre.

"No, I don't," said Charles. "I think that, as telescopes get better, people will find some that we missed. Not to mention that we can only see the night sky up here in Europe. The view from the southern part of the globe is surely different."

"Okay," said Pierre, "M45 is logged in. If you like, I'll go talk to the printers tomorrow."

"Researchers kept adding to the list," Albert said. "Messier's own last catalog was published around 1781, and those objects are still known today by their Messier numbers."

"I see what you were trying to show me," I said to Albert. "Humans were gradually coming to realize that the cosmos was much bigger than they thought."

"Exactly," Albert said. "They were also wondering what space itself was."

SNIFFING THE ETHER

After one look at this planet, any visitor from outer space would say, "I want to see the manager."

—William S. Burroughs

"THIS BUSINESS ABOUT WHAT SPACE IS," I said. "I must say I find that fascinating. What can you tell me about it, Albert?"

"Where to begin?" he mumbled. "I suppose I could summon up some mist, and we could go back to see Isaac again."

"Newton?" I said. "What did he have to say about it?"

"He speculated that there could be an ether throughout space," Albert said. "That it was denser near large planetary bodies, and rarer in the space between them."

"I see," I said.

"Or we could take a look at another man whose ideas I quite liked," Albert said. "That was Joseph Larmor. He grew up in Ireland, ended up teaching in Cambridge, and published a book called *Aether and Matter*, around 1900.

"The basic idea, as I think you know, was that matter and space are somehow pervaded by what they were calling *ether*. This is different, of course, from the ancient idea of the ether as a kind of spiritual place inhabited by gods. You know, it might be fun for us to pop in on another ether thinker, James Clerk Maxwell. You'd like that, because he was a Scot, like you."

"How did you know I'm Scottish?" I asked.

"You say so in one of the books you wrote," Albert said. "Now, mist!"

When it cleared, we were looking at a man with a magnificent and rather unruly beard. He was hunched over a desk, writing carefully while occasionally lifting his head to think.

"He was brilliant, you know," said Einstein. "There was a competition to figure out what the rings of Saturn were. Working in Aberdeen in Scotland, Maxwell devised the mathematics to show that a solid ring would wobble, that a liquid ring would break up, and that the best solution was a ring made up of particles.

"He was the only entrant who dealt adequately with the problem, so he won the competition. Then, in the 1980s, satellites were able to take pictures that confirmed Maxwell's theory."

"I'm thrilled to see Maxwell at his desk," I said. "Although it's a bit boring just watching him writing."

"That's how theoretical physics works," Albert said. "It's lots of thinking, scribbling, erasing. Ninety-nine-point nine percent frustration, and point one percent breakthroughs. It's not a spectator sport, you know.

"Back to the ether. Maxwell thought that there might be a mechanical explanation for the cosmos. Perhaps a medium that existed throughout space, with each particle of energy causing a knock-on effect to the next."

"I have a lot of sympathy for that idea," I said.

"I have to confess to having derailed it a bit," Albert said. "Maxwell was followed by others working on the idea, but eventually it was realized that my own work could explain events without the need for any ether at all."

"So you're to blame," I said.

"Guilty as charged," Albert said, with a smile. "But, tell me, why are you a fan of the ether?"

"I recall something from my youth," I said. "I was in high school, in a science class of some kind, but I wasn't terribly interested in the beakers and Bunsen burners all around me. Then the teacher showed us a vacuum jar. It was very heavy and was fitted with an extraction tube and a pump. He went on to suck out the air inside, and then proclaimed that the jar contained a vacuum. 'It's absolutely empty, boys. Just like in outer space,' he said."

"This didn't pique your interest in science?" asked Albert, intrigued.

"I suppose it could have," I said, "but I was kind of a lost soul in those years. Had no sense of purpose in life."

"Perhaps your teachers should have inspired you more," Albert said.

"I've always thought that it wasn't really their responsibility to do that. It was mine. Anyway, what the teacher said about outer space resonated with some of the pupils. I remember them talking with wide eyes about what it would be like to go into space. What the teacher could have done more usefully, however, is tell us that the jar was not really empty at all."

"Ah," Albert said, with a smile, "you're going to talk about *vacuum energy*?"

"Right," I said, nodding. "I learned much later that scientists have known for around a century that a vacuum isn't really empty. Instead, it contains an energy field. What's intriguing is how colossal the value of the energy is. It's mind-blowing. What I don't understand is this: why don't we hear much more about that?"

Albert nodded. "Physicists are, of course, aware of the concept of vacuum energy. But you'll find them arguing furiously about whether it has any meaning in the grand scheme of things."

"Very well," I said. "I think I've finished with the ether. My head is spinning. I think I'll need a Martini to calm it down."

STOP EXPANDING!

There is just one thing I can promise you about the outer space program—your tax dollar will go further.

—Werner von Braun

I WATCHED AS ALBERT PREPARED to summon up his mist again.

"Can't you do this in some different colors?" I asked, testily. "This gray is awfully boring."

"You're just annoyed that I keep insulting you, aren't you?" he said.

"I'll get my revenge," I said. "You're going to be impressed when I unveil my brilliant ideas."

"While the world waits breathlessly," Albert said, "let's visit a real genius."

I found myself looking at a man seated at a very large telescope.

"That's Edwin Hubble," Albert said. "He made many very important discoveries, and whatever weird theories you're developing will have to take those into account.

"As an astronomer, he was somewhat fortunate, because he arrived at the Mount Wilson Observatory in California in 1919. It was only a few years later that they activated the Hooker telescope, the world's largest at the time.

"Hubble's father, by the way, had wanted him to study law, and Edwin complied, going first to Chicago then later to Oxford in

England. While there, he had to take a few scientific courses, and he got addicted.

"After his father passed away, Hubble did post-grad studies in Chicago, and that's how the world got its most famous astronomer. After Galileo, of course."

"Sounds like he was lucky to pursue his interest," I said. "Not just that, but to be was in the right place at the right time. With that huge telescope, he had the best possible equipment to snoop around in the night skies."

"Right," Albert said. "He made some remarkable deductions. For example, he was studying what they call nebulae, and folks back then thought those were just wispy clouds of stuff floating around in our galaxy. But Hubble showed that they were galaxies in themselves, outside our own Milky Way."

"That must have surprised the scientific community," I said.

"Not really," Albert said. "Back in the eighteenth century, Kant had theorized that there might be such galaxies, but the scientific community had fiercely resisted the idea, right up to Hubble's time."

"Why do they do that?" I asked. "Why do scientists resist new ideas that come along?"

Albert smiled. "Because new ideas are a threat. If you think about it, scientists have carefully worked their way into their own preferred communities of belief, with their own positions and financial support of one kind or another. They resist anything that might change that."

"I thought scientists were supposed to be interested in the truth," I said.

"They do say that, don't they?" Albert said. "But here's what happens to a revolutionary new idea. First, it's derided as being nonsense. Then, when it looks like interest is picking up, it's attacked viciously; not just the concept itself, but the proponents of it."

"What if the idea is correct?"

"Then it is eventually accepted as being obvious," Albert said, grinning. "Not so much by the old guard, though. More often by the next generation of scientists that grows up used to the new idea. It was Schopenhauer who said all that."

"That's sad," I said.

"Then let me cheer you up," Albert said. "Hubble's idea of galaxies outside our own was eventually accepted, and he went on to look at the red shift of those galaxies."

"What does that mean?" I asked. "Hubble saw red light coming from distant stars?"

"Oh dear, do I have to explain everything?" Albert said. "Physicists talk of redshift when radiation from a moving object shifts towards the red end of the color spectrum. You don't have to be seeing red necessarily. It's a bit like the sound you hear when a train passes."

"You mean the Doppler effect? Yes, I know about that."

Albert nodded. "The red shift puzzled Hubble, and he really didn't know what to make of it. It was eventually a priest who figured out what it meant. He said that it showed the universe was expanding. Hubble never quite accepted that."

"A priest?" I said. "What's a priest doing messing with science?"

"You display your ignorance yet again," Albert said. "The Church has always had an interest in science. In fact, for many centuries, science was practiced only in monasteries. They wanted to understand God's world, you see. His magnificent creation. And the priest I'm thinking of was doing that kind of research in the twentieth century. He lived and worked in Belgium, by the way."

"Interesting," I said. "I lived there for many years."

"And yet you've never heard of the famous priest, Lemaître?"

"No," I replied.

"What were you doing all that time?" Albert said. "Sleepwalking through life?"

"Maybe I was," I said, feeling a little dejected

A COSMIC EGG?

In the beginning, the universe was created. This has made a lot of people very angry, and has been widely regarded as a bad move.

—Douglas Adams

OUR NEXT STOP, ALBERT SAID, "is a city called Louvain, not far from Brussels. You said you lived in Belgium for a time, so I hope you've heard of it."

"Indeed I have," I said, smiling. "I took a job in Brussels in the mid-1970s, leaving England behind. A friend from London had done a similar thing, accepting a job in Louvain. We met a couple of times for dinner, and I'd drive back home without giving that town a second thought."

"You shouldn't have been so dozy," Albert said. "The city has a fascinating history. For our purposes, however, I'm mostly interested in the Catholic University of Louvain, founded in the fifteenth century. And in one particular man who conducted research there—a priest by the name of Georges Henri Joseph Édouard Lemaître.

"In the late 1940s, he got the entire Catholic Church very excited. Let's listen in as Pope Pius XII discussed the Belgian's ideas with some colleagues."

The mist swirled, and I found myself inside the Vatican, watching a group of robed men holding a meeting.

"Good morning, Your Holiness," said Sarducci. "Thank you for joining our team here today."

"Good morning, everyone," said the Pontiff. "First, let me thank you for all your work. It can't have been easy, trying to follow the complexities in Lemaître's thinking."

"Your Holiness," said Sarducci, "did you manage to read the technical summary we sent to you yesterday?"

"I skimmed it," came the reply, "but, you know, I'm less interested in the physics of it. I'm focused more on the implications for our Church."

"I see," said Sarducci.

"What happens if Lemaître is right?" asked the Pope. "What should the Church do about this matter?"

"We have given this some thought, Your Holiness," Sarducci said. "At times during our work, we couldn't help thinking back to how the Galileo situation was managed centuries ago. We're not sure if this new issue is as significant as that, but we did wonder.

"Our Church pushed back against Galileo, because the idea of geocentrism seemed to contradict scripture. In Genesis, the earth is created first, and then the sun and moon. At various points in Psalms and 1 Chronicles, we can read that the Earth does not move—"

The Pope interrupted. "How helpful of you, Sarducci, to tell me what our scriptures say."

"Sorry Your Holiness," Sarducci said, his voice nervous, "I wasn't implying that you didn't know . . ."

"Relax," said the Pontiff, with a smile, "I was just yanking on your rosary." He paused, then said, "Let's cut to the chase. I'd like to get ahead of the game this time. I want to make a statement. Something like . . . we see the revelation provided by God to Lemaître as proof of the Creation."

"And if Lemaître is eventually proved wrong, Your Holiness?"

"No harm, no foul," said the Pope. "In that case, we just revert to the Creation as it was always understood, and the faithful will forget what we said about—what was it, again—this primeval atom? Do we all agree?"

Sarducci nodded. "I would be most hesitant to argue with Your Holiness."

The Pope smiled. "Yes. One of the perks of being infallible is that folks usually end up agreeing with me."

Albert smiled. "That meeting came to an end, and Sarducci went off to have some well-earned Camparis at his favorite bar on the Via Veneto. Of course, some of the staff stayed behind, to draft the Pope's statement supporting Lemaître. You might think, then, that all was well that ended well, but Lemaître wasn't as pleased as you might think. Let's watch him."

We were hovering over a modest-looking man hunched over some papers on a desk. While I was gazing at him, a door opened and a young man entered.

"Father Lemaître," said the young man, holding up some papers, "our office received a communication about the meeting with the Holy Father. Sarducci sent us a summary of their conversation."

"I heard, Robert," said Lemaître, "although I rather wish that I hadn't."

"Why not?" asked Robert. "Isn't it a good thing to have the Pontiff focusing closely on our work here?"

"You'd think so, wouldn't you?" sighed Lemaître, "but the Church always gets itself into difficulties when it messes with science. Please give me the full report, so I can read it for myself."

The priest took the pages from Robert and started leafing through them.

"Yes," said Lemaître, "this is what I was afraid of. His Holiness doesn't care much about the physics of our work. He probably didn't even understand the technical passages. He seems to be keen to rush out and

declare that our results are a confirmation of God's work—of The Creation, in fact."

"Well, I can understand that," said Robert, "your work here has shown that there was—how did you put it—a primeval atom, or cosmic egg, from which everything began."

"That's where the science leads us," said Lemaître. "Also to theorize that there may have been a colossal explosion to scatter the initial material into what we know as the universe."

Albert sighed. "I was skeptical of this myself, because it made some of my earlier work look a little silly. I insulted him. I was wrong, and it was painful to admit that, but I have to give credit to Lemaître. He solved one of the biggest mysteries about the origins of the cosmos."

"Where was he educated?" I asked.

"Lemaître had spent a year in Cambridge," Albert said, "working with the famous researcher Eddington. Then the priest spent a year in Harvard. You can't get much better creds than those, can you?

"He did, however, hit a problem with me. I had favored the *steady state* theory, the idea being that the cosmos wasn't expanding, but was somehow stable. I was indeed wrong, and I eventually swung round to Lemaître's way of thinking, giving the Belgian priest great praise for his discovery.

"Around the same period, Hubble was doing his thing, and he added to our amazement by showing that the constellations are actually accelerating away from each other."

"So," I said, "more proof for Lemaître."

Al nodded. "This was seen as further support for the cosmic egg and Big Bang theories. Its only problem, however, was that it couldn't explain the very beginning of the universe.

"Regular folks had a very simple, and understandable, question: if the universe exploded from a cosmic egg or primeval atom, what was in place before that happened? Stands to reason, they'd say, that there must have been *something* there before the egg?"

I laughed. "Like some kind of space chicken?"

Al chuckled. "Unfortunately, although there are theories, nobody has ever been able to prove how the universe began."

I looked at Albert. "Well, hold on to your socks, my friend, because I think I know. It's all starting to come together in my mind."

"Oh dear," sighed Al. "Forgive me if I don't hold my breath."

"You've been mentioning your own work," I said. "Tell me more."

"Anything to help your education," Albert said.

PATENTLY BRILLIANT

The world, we are told, was made especially for man—a presumption not supported by the facts.

—John Muir

"THIS CALLS FOR A SPECIAL dose of mist." Al twirled. When everything cleared, we were in a room where there was a desk, on which folders and papers were scattered.

"This is my office," Albert said. "At the precise moment when I died, in 1955."

"I feel sad, seeing this," I mumbled. "It looks like you were still working on something. Those chalk marks on the board, your books lying open. And the desk is—I'm sorry for saying this—a bit of a mess."

"I was always working on something or other," Albert said, quietly. "In fact, I was scribbling down equations in an old notepad just a few hours before my death. And my desk was always disorganized like that. Besides, I wasn't expecting to die that day."

I tried to lighten the mood. "You know what they say, don't you? An untidy desk is a sign of an untidy mind?"

Albert chuckled. "Then do tell, what is an empty desk a sign of?"

I laughed.

"Don't mind me," Albert said, "I'm just messing with words. I used to enjoy doing that in English. I'd say things like . . . if crime fighters are fighting crime, then what are freedom fighters fighting?"

"That's a good one too," I said, smiling at him.

"You know, my later years in Princeton were rather quiet," Albert said. "Students all over the world would compete like crazy to study here at Princeton, just because I was around. Then, when they got here, they were usually too shy or intimidated to talk to me.

"What they didn't know was that, if they had walked up to my front door on Mercer Street and rung the bell, I'd have invited them in for tea, to talk physics and swap jokes. But those young folks were in awe of what was supposedly the greatest brain the human race ever had."

"When did your celebrity begin?" I asked.

"I tried to avoid the fuss, you know," he answered. "At the height of my fame, people would stop me in the street, and ask me to explain . . . *that theory*, they said. I would reply that I was always being mistaken for Einstein, then I'd walk on.

"To answer your question, however, I'd have to say people started to pay attention around 1905. You probably know that I was still working as a patent clerk when I published papers that caused a bit of a stir in the physics world?"

"Yes, I'd read that. It seems that the good thing about government work is that it lets you be your own boss."

Albert laughed. "You're right, I had been somewhat underemployed dealing with those patents, so it gave me time to think quietly, sitting there in my office. One of my desk drawers contained all my scribblings, so I called it my Theoretical Physics Department.

"When I would daydream, it wouldn't be about the things that preoccupy most other people. For example, a particular issue that troubled me was this: what would it be like to chase a beam of light?

"Then there was that incident when a local man working on a roof fell to the ground. Thankfully, his fall was broken by bushes, so he survived. But he described having had a sensation of weightlessness as he fell, and that interested me."

"I've occasionally felt that in an elevator," I said. "At the moment when it starts to move down quickly."

"Exactly," Albert said. "I puzzled a great deal over what people would feel in elevators, rockets and trains. Also, what would happen to a beam of light entering a side window of a speeding car or train? How would it cross the carriage? Would it go out the window exactly opposite, or would it be deflected?"

"I'm not sure I could focus on issues like that for long," I said. "I'd worry that I might go crazy."

Albert laughed. "Lots of physicists do go crazy." His face turned dark. "They even kill themselves. Like that poor man Boltzmann. He hanged himself when he couldn't take any more." Albert smiled gently. "Still, they put his most famous equation on his tombstone."

"The one on entropy?" I asked.

Albert laughed. "You *have* been doing some reading. I'm impressed."

"Luckily," I said, "you managed to figure out enough to stay sane. "That year of yours, 1905, what did they call it again?"

"My *annus mirabilis*," said Einstein. "Meaning *my wonderful year*. A needless use of the Latin. We have enough confusion in the world without that, don't you think?"

"What was special about that year?"

"I produced four papers," Albert said. "The titles are a bit labored, as is often the case with physics articles. The first one was called *On a Heuristic Viewpoint Concerning the Production and Transformation of Light*. It was in German, of course, and that made it sound much more impressive. It dealt with luminous energy being transmitted only in specific quantities. To my horror, that thinking helped others develop the quantum theory.

"The second paper, *On the Motion of Small Particles Suspended in a Stationary Liquid, as Required by the Molecular Kinetic Theory of Heat*, was helpful in demonstrating the existence of the atom." Albert looked at me. "At that time, you see, many doubted that the atom even existed.

"My third paper, *On the Electrodynamics of Moving Bodies*, claimed to show that the speed of light was independent of the motion of an observer. This was the idea that became known as the *Special Theory of Relativity*.

"And the fourth paper, *Does the Inertia of a Body Depend Upon Its Energy Content*, was the one that gave humanity its favorite equation."

"Even I know that one," I said. "It's $E = mc^2$, meaning: energy equals mass times the speed of light squared."

Albert laughed. "It's odd," he said, "My paper didn't have the equation quite like that. You see, I used the letter *L* for energy, not *E*. And I used *V* for the speed of light, not the lower-case *c*. Tell me," Albert went on, "what do you think of when you read that equation?"

"Hmm, it tells me that a huge amount of energy can be generated by a small amount of mass. That's because the speed of light squared is a massive multiplier."

"Right," Albert said. "To my distress, this led to the development of the atomic bomb."

I smiled. "I also think of the equation another way. If you turn it around and divide both sides of it by the speed of light squared, you get $m = E/c^2$, showing that mass is really a form of energy."

"Why did you need to turn it around and fiddle with it?" Albert said, frowning. "Wasn't that already obvious?"

"It just made it clearer to my lesser mind," I said. "I also happen to think that this is hugely important. I suspect that, when I finally disclose my ToE, this will figure prominently."

“Yes, your brilliant Theory of Everything, explaining the cosmos to us all. I can hardly wait.”

“You became very famous,” I said. “Not just for scientists, but with the public.”

Albert chuckled. “I became friends with all sorts of famous people, like George Bernard Shaw.”

“Oh yes,” I said, “He was a brilliant writer, just like me.”

Albert looked sadly at me. “I was also befriended by the Queen of the Belgians. A nice lady. Then there was Charles Chaplin, who invited me to the premiere of his movie *City Lights*, and the crowd went crazy when we arrived.

“Charles turned to me and said something like, ‘They’re applauding me because they understand what I do. They’re applauding you, Albert, because they have no idea what you do.’”

“Very funny,” I said.

“He was a brilliant man, Chaplin,” Albert said. “A fine composer too. I really like his song *Smile*. You know the one?” Al started to sing. “*Smile, though your heart is aching, smile when your heart is breaking . . .*”

“Yes, I know it,” I said. “Wonderful.”

“You never know,” Albert said, “maybe when you publish your ToE, the world will applaud you like that.”

“Because they don’t understand what I’m doing?”

“No,” Albert said, with an evil smile, “because *you* don’t.”

I didn’t mind the insult.

It’s not everyone who can say that Albert Einstein sang to them.

WAS THAT A BANG?

The theory in Hiroshima is that the atomic heat was so great that they burned instantly to ashes.
Except that there were no ashes.

—Wilfred Graham Burchett

WHO DOESN'T LOVE AN EXPLOSION? Most of us do I think, and I say that because every year, in the days that lead up to July 4th, thousands of New Jersey residents cross the border into Pennsylvania.

They're cruising along the highways, making for temporary tented stores where you can buy fireworks. You see, they're not sold to the public in New Jersey, but are easily available over the border in PA.

A fireworks-hunting day is seen as jolly good fun; makes the folks from New Jersey feel like romantic old-time smugglers, heading home with contraband in the trunks of their cars.

I live in rural Pennsylvania and, on July 4 every year, some of my near neighbors (I'm never exactly sure who it is doing this) like to set off rockets and all sorts of whizzing and whistling projectiles that soar into the sky.

By the way, New Year's Eve here is a little different stylistically. At midnight, some of the locals like to step outside their homes (I do hope it's outside) and fire off their shotguns. For those of you unaware of this, it's called *celebratory gunfire*, a fine American tradition.

"So, Al," I said, "we humans love bangs, and the bigger the better."

"I agree," he said. "Nature does a pretty good job in this department. Just think of the volcanic eruption at Krakatoa, near Java. I was four years old at the time, and two thirds of that island, perhaps around nine or ten square miles of it, just disappeared off the map.

"The noise of the explosion was heard three thousand miles away, and the tsunami was still being felt several days later. As many as 100,000 people may have died."

"Impressive, right?" I asked.

"Not even close," said Al. "Nature had already done much better with an eruption around the year 1626 BCE, at a location known as Thera, in the Eastern Mediterranean. But don't reach for your atlas, Ian, since much of Thera no longer exists, except for the part of the island now called Santorini.

"The Thera bang may have been a hundred times bigger than Krakatoa. Most of the island simply shot into the air, and the pyrotechnics involved must have been spectacular."

"I know a bit about that one, Al." I said. "I wrote about Thera in my book on religion, in which I outlined the theory that this particular eruption may have been responsible for the Hebrew legends of the Exodus. The good book says that six hundred thousand of those poor souls, frightened out of their wits, anxiously sought safety as they headed to the land they called Canaan."

I looked at him. "Okay, so we're done right? There can't be anything bigger than that in the explosion department."

Al frowned. "We don't really know. Researchers aren't sure which Earth-based explosive event was the most dramatic of all time. They might, however, wonder about the meteor that may have crashed into our planet and sent up so much dust and debris that the world's temperatures plummeted. Thereby helping to kill off our friends the dinosaurs—even the purple ones.

"We've also been a bit unkind to humans. We can do much better than just having little fireworks displays. Better even than the Macy's July 4th extravaganza along the East River in Manhattan. You see, we can make a pretty big noise when we kill people.

"In 1945, we had been bombing Tokyo and other cities in Japan and had already caused hundreds of thousands of deaths. But the Japanese hadn't surrendered quickly enough so, on August 6th, the B-29 bomber known as *Enola Gay* flew over Hiroshima and dropped the bomb that would immediately kill 80,000 residents of the city."

I nodded. "I visit Washington, DC from time to time. I've taken a peek at the Enola Gay. It's in the National Air & Space Museum."

Al continued his story. "A few days after Hiroshima, when the Japanese authorities still hadn't caved, another group of B-29s flew over Japan, working their way down a list of targets.

"There were several cities they could have chosen for the second bomb, but those were covered in clouds. It was the sunny weather and clear skies over the city of Nagasaki that doomed the people there.

"The crew in one of the accompanying B-29s told of looking back at the explosion. They saw a giant ball of fire that turned into a seething purple pillar. After about forty-five seconds had elapsed, they could look back and see that the explosion cloud had reached their cruising altitude, at around 17,000 feet."

"Interesting, Al," I mumbled. "So, now that we've thought about those mighty events, how about the so-called Big Bang? How big was that one? And, more importantly, how and why did that happen?"

THE BANG

As in an explosion, I would erupt with all the wonderful things
I saw and understood in this world.

—Boris Pasternak

I GET IT, ALBERT, EXPLOSIONS ARE FUN, but I'm only interested in the first one of all, the one they call the Big Bang."

"How do you know it was the first?" Al said. "There could have been a whole series of them, with the cosmos collapsing back onto itself, and then exploding again and again?"

"Hadn't thought of that," I said.

"Real scientists," Al said, "who, unlike you, do think deeply about such things, have been working hard, trying to move the clock back to find out what happened in the very first instants of the Bang. It's kind of like peeling an onion, you know."

"The priest Lemaître was wrong?" I said.

"What?"

"It wasn't a cosmic egg, it was a cosmic onion?" I chuckled.

Al looked disgusted. "Have I told you that you're a moron?"

"You have," I said. "So, cough it up. What do the nerds say?"

Al turned serious again. "They spend a lot of their time trying to figure out when the first nuclei formed. Scientists can't really say what happened in the very first instants of the explosion, but they think it was initially too hot for any structures to take shape. After a tiny fraction of a second, however, some of the sub-atomic components we know would have appeared."

"No elements had formed yet?" I asked.

"Exactly. It was just too darned fierce. Those simpler components were being destroyed as fast as they took shape. You only need energy of a few million volts to break up a nucleus, and the explosion was colossally more powerful than that."

"Was everything in balance?" I asked. "The same number of positive charges as negative?"

Al smiled. "That's the most perceptive question you've asked me during our travels. I must be having a great influence on you, dear boy. So, why exactly did that occur to you?"

"I have a hunch that it's hugely important," I said. "But don't let me stop you. Go on with the slow-motion Bang."

"As the next few seconds of the explosion occurred," Al said, "it was still too hot for atomic nuclei to form for any noticeable length of time. As time continued to tick away, nuclei were almost taking shape, but were still quickly being broken up."

"When did it stabilize?"

"Depends what that means," Al said. He looked serious again, as if we were getting close to something meaningful. "You see, the important thing for me is when the material appeared that would give us our stars.

"The components of that process would probably have been there in the first few minutes. There would have been formation of helium, deuterium, and hydrogen nuclei. Hydrogen, by the way, is still the most common element in our universe."

I looked at my friend. "Al, there's something I've always wondered about. If you have a hydrogen atom, with one electron whizzing around the nucleus, then why doesn't that electron crash down into the proton and annihilate?"

Al chuckled. "Yes, I told you, didn't I, that Rutherford wondered the same thing. What do *you* think?"

"I'll guess that the underlying energy field of the cosmos—that vacuum energy we spoke of earlier—is pervasive. Imagine the entire cosmos filled with energy. Little positive and negative charges, just twinkling into existence."

"You're talking about dark energy, aren't you, Ian?"

"I am. And the interesting thing about fields of energy like that is that the plus charges meet the negatives, and then they do something very interesting."

Al nodded. "They spin."

"Right," I said. "So, just think of that. An entire cosmos filled with energy that pulses into existence, then is pushed aside by the next charges that appear. The net effect being that everything is pushing and twisting furiously. That might be why the electron doesn't crash down. There's too much energy getting in its way."

Albert looked at me. "Remember, too, that the forces holding the nucleus in place are strong but are localized. That allows electrons to flit around in their orbits, and it's why elements can exist in the first place."

I looked at him. "We're getting somewhere with all this, Al. You see, I'm also thinking what happens when you play with two magnets and some iron filings on a table. If you move them around just right, you'll get spirals forming in the filings."

Al smiled. "I suppose you think that it was this activity—the spiraling of the energy acting on the hydrogen atoms—that gave us our stars?"

"I do," I replied. "Hydrogen being spun into clouds, then those areas being concentrated into supernovae where denser elements formed. The supernovae blasting that material into space, where it could re-form into planets like ours."

"Again, these thoughts of yours are not new," Al said. "Carl Sagan said that every atom of iron in our blood came from a supernova somewhere. In fact, he said we are all made of stars."

I smiled. "True, but this has an interesting implication for me. You see, if everything is being pushed all over the place, then is there really gravity as people think of it?"

Al looked shocked. "Now hold on there, you whippersnapper. I am the defender of gravity. Haven't you seen those demos where people put a bowling ball on top of a bed, then roll billiard balls around it to show how the attraction works?"

"I have, Al, and they're great entertainment. But I still prefer to think that there is no such thing as a pulling gravity force. Things in the cosmos only move when they're pushed. If it appears that they're being pulled, it just means that it looks that way. It's all to do with exactly *how* they're being pushed.

"Don't you remember, Al? Earlier, we were saying how sad it was that it took so long for people to wonder whether it was the Earth moving round the sun, not the other way?"

Al went quiet.

"What's wrong?" I asked.

"Oh, nothing," he said. "But I don't feel so enthusiastic about helping you anymore."

"Well, if you want a break, then that's okay. Besides, Al, I'm going on to dig into the issue of life. Maybe you're not up to that challenge?"

There was a pause, then Albert let out a disgusted grunt.

"Our first stop," he said, "will be to look in on Diocles. You'll need to know about him, and some of the other great thinkers. God, I just hope the other Time Lords don't find out that I'm ferrying a moron around through the centuries."

"That's the spirit," I said. "Now, do the mist thing."

LIFE

Is there any such thing?

DIOCLES

No man should marry until he has studied anatomy and dissected at least one woman.

—Honoré de Balzac

ALBERT," I SAID, ARE YOU SURE you're up to this? This isn't your science stuff. This is about how life came to be."

"Listen, young man," Albert said, "You seem to be getting more and more pushy with your own thoughts. But don't forget that I know everything about everything, and only I can do the twirly-finger thing and take you back in time.

"As I said, we're going back right now to around 325 BCE, to visit a spot that was known as Carystus. I'd like you to watch a man named Diocles in action."

"What's Carystus called now?" I asked.

"Karystos, what else?" Albert said. "Now, pay attention."

When the mist cleared, we were looking down at a man who was hunched over what looked like the mutilated body of a deer. I grimaced at the sight of the blood and the fleshy matter that was lying around.

"Help me clean all this up, Akakios," Diocles said to the young man at his side.

"Did you learn anything from this animal, Master?" Akakios asked.

"Nothing new," Diocles said. "I just confirmed the general layout of the organs, the muscles and the bone structure. It was quite similar to some of the other quadrupeds I've dissected."

The young man dragged away the carcass, then started mopping up the blood.

"If you don't mind me saying, Master, it's lucky you are a respected doctor in these parts, and that you help people with their ailments. There are those who frown at your apparent delight in killing and slicing up animals."

Diocles smiled. "You mark my words, Akakios, some day this will be recognized as important work. We desperately need this knowledge in order to save and cure our patients."

Diocles turned to a side table and picked up a small instrument.

"See this, Akakios? It was because of my work on corpses that I was able to design this tool. It's excellent for removing arrowheads that are lodged in soldiers' bodies." He placed the object back on the table. "Using it, I can minimize damage to surrounding tissue."

Akakios looked at Diocles. "What now, Master?"

"I think, Akakios, that I have done enough research to be able to publish my findings in a book that discusses how bodies are structured."

"This could almost be the beginning of a new science, Master."

"I've been thinking the same thing. I'll call it . . . Anatomy."

"Is that true?" I asked Albert. "Did he develop the science of anatomy?"

Albert smiled. "He may have been the first to employ the term. But you have to realize that so-called doctors had been trying to puzzle out life forms for centuries, and even longer.

"They would kill and dissect animals to see what lay inside. And some of the more ruthless researchers didn't hesitate to kill and examine prisoners of war."

"Gruesome," I said.

"Aristotle was also one who built a classification system for the types of animals that he studied, and he saw a guiding hand behind it all. For example, he thought that animals had been given tusks or a horn, but not both. He also noted that cattle had two stomachs, and felt that a designer had added the second one to compensate for weaknesses in the first."

"I see," I said. "So, no hint of spontaneous evolution in their minds back then."

"Right," Al said. "Let's speed things up here and talk about just that."

"So, we're going to see Darwin? Great."

"Yes, we will," Al replied. "But first, we should look at a couple of Frenchmen who can set the scene for us."

BONEY & THE BARON

History is a set of lies agreed upon.

—Napoleon Bonaparte

PARIS AGAIN,' I SAID, "I'm always happy to revisit this town. But which era we in now?"

"The Napoleonic one," Al said. "In fact, here he comes now."

As the mist cleared, we were looking down on a rather splendid room, in some kind of palace. Gilded chairs surrounded an ornate table, and impeccably-dressed waiters stood by the doors and walls.

The doors opened and, to my amazement, there was the Emperor striding in, and looking just like his portraits. Short, as I expected, but with a very confident bearing.

There was only one other person in the room who wasn't dressed like a servant. He rose to his feet as Napoleon approached.

Napoleon smiled at the man.

"My dear Baron Cuvier, welcome to the palace. I've really been looking forward to our meeting, because I find you a most interesting person indeed."

"Thank you, Your Imperial Majesty, it's a pleasure to meet you, and to do so here in the Louvre. I was hoping that you were not upset at my inability to accompany you to Egypt? I was so busy at the time."

"No problem at all, dear Baron."

"How fitting that you should live in these fine accommodations," Cuvier said, looking around at the opulent surroundings.

"I'm just following in Louis XIV's footsteps, you know," said Napoleon. "I thought I'd move in because they've just renovated the place. Now, please tell me about your latest thinking on the natural order of things. Some are saying that you no longer believe that the animal world is unchangeable."

"Quite so, Majesty. As you know, I had always assumed that no animal species can ever have disappeared from the Earth. The idea being that God would not put an end to any of his creations."

Napoleon nodded. "I held that opinion myself, although I confess to not having thought much about this until I learned of your work."

The Emperor then motioned to the chairs, and the two men sat down then waited until the servants approached to pour golden liquid into two glasses.

"It's brandy," Napoleon said. "Made at the House of Courvoisier and aged for years in oak barrels. I enjoy it so much they're thinking of naming it after me."

They both laughed, then sipped at their drinks.

"Your Imperial Majesty," Cuvier said, his voice hesitant, "I am grateful for today's meeting. You see, I am aware that I have created quite a fuss among the country's religious leaders."

"Indeed you have, Cuvier. They do not appreciate your claim that animal species can become extinct, but I now think that this is unavoidable."

"I feel a certain sadness," Cuvier said. "Some in the Church call me a heretic, but I am a believer in our God. I have simply come to the conclusion that He caused not just one flood, but a series that killed many animal species, laying their bones down in different layers in the land."

"Is this true?" I asked Albert. "Not the bit about God and floods. But that Cuvier changed the world's thinking about extinction?"

Al smiled. "He did. And he set the stage for someone else to come along and figure out the next part of the grand puzzle. Better get your seasick pills. We're going on a sea cruise."

"Where to?"

"The Galapagos Islands."

CAPTAIN'S LOG, GALAPAGOS

I have called this principle, by which each slight variation, if useful, is preserved, by the term of Natural Selection.

—Charles Darwin

IT CERTAINLY LOOKS IDYLLIC, ALBERT," I said, looking down at the Beagle as it floated gently near the islands.

"I agree. Right now, the Captain—a man named FitzRoy—is having a meeting with a certain Charles Darwin. They've been doing this frequently, on their journey all the way here from England.

"Can we listen in?"

"Of course," Albert replied, and he waved his hand.

I found myself inside one of the cabins, looking at the two men. Darwin was immediately recognizable from photos that I'd seen. There was his full beard, and his rather unpleasant bone structure.

"My dear Darwin," said FitzRoy, "how much longer do you think we need to stay at these islands?"

"Why, Captain? Do you find them unpleasant?"

"Not particularly. But they are less than excessively diverting. The men think so too, they're tired of eating turtle, and they complain that there are

no alehouses here. They also whine incessantly because there are no women."

"Ah yes," Darwin said, smiling as he nodded. "That particular curse of a sailor's life."

FitzRoy frowned. "Somehow, they make do. But I shudder to think about what goes on below decks here at night."

There was a short silence, then Darwin looked at FitzRoy.

"Dear Captain, I ought to let you know that I am becoming more and more certain of those ideas I explained to you earlier in our voyage."

"This idea of a selection process?"

"Indeed. One fine illustration of the concept is the distribution of geospizas on the islands. My servant Covington has been collecting them for me, and they have some attributes that intrigue me."

"Those little birds? How so?"

"Well, Captain, the most striking feature is the gradations in the size of their beaks. I've drawn the conclusion that they have become finely adapted to the food sources on their particular islands. I also suspect that these birds are only to be found here, in the Galapagos."

FitzRoy frowned. "But is that variability not simply a demonstration of our good Lord at work? How he carefully designed each of his creations to suit his purpose?"

Darwin chuckled. "I can't help thinking that if I were a divine creator, I would be sorely tempted to save myself a great deal of very detailed work. Instead of designing each bird, cat, and reptile so carefully, I would set in motion a method that would result in animals changing their features depending on their environment, and on how it changes over time."

"But how could that possibly work?' FitzRoy asked.

"Tell me, FitzRoy, do you have children?"

"No, Sir, not yet. I do plan to marry at the end of this long voyage. Although I fear it will have been a journey of four or five years, and not the two years we originally intended."

Darwin nodded. "If you go on to have, say, four sons, do you imagine that they would be identical?"

"Of course not," FitzRoy said. "One son might be fast, another slow, yet another strong, and so on. That is the way of things."

"I agree," Darwin said. "Consider, then, what might happen if the survival of your sons, early in their lives, depended on their being fleet of foot. What then?"

FitzRoy looked at Darwin. "I dare say that only the fast boy would live."

"Agreed," Darwin said, chuckling. "Only he would reach the age where he could find a mate and create his own family. So, his swiftness may be somewhat more likely to be passed down to his children."

"I see what you mean."

Darwin nodded. "I call this process . . . natural selection."

"I understand what you say, my dear man," FitzRoy said, "but how on earth does this variability happen? What makes it that my eventual wife and myself would create children with such different traits?"

Darwin sighed. "That, is a question for another mind to unravel, I fear."

I turned to Albert. "I've heard that, even today, about forty percent of Americans prefer to believe that it was God who created life. They don't care for the ideas of our friend Darwin."

"You may be right," Albert said. "We do seem to be rather clueless in some ways."

"It's a pity," I said, "that nobody has ever been able to prove Darwin's theory."

"Maybe they have. Let's go see."

WORKING IN THE LAB

If evolution really works, how come mothers only have two hands?

—Milton Berle

ALBERT, I'M ALWAYS SURPRISED to find people who adamantly oppose the idea of natural selection."

"Those pesky creationists," Al said, with a smile. "I know what you mean. It wouldn't matter how much proof you laid in front of them, they would just keep the blinders nice and tight over their eyes."

"Proof?" I said. "Isn't that the problem? Those folks say that evolution is an interesting idea, but there's never been any proof of it whatsoever."

Albert nodded. "It's hard to provide proof for something that takes so long to run its course. Although there have been instances noted here and there, like moths that quickly developed a different color on their wings, to adapt to sudden changes in their area's foliage.

"There's also the annoyance of the Cambrian Explosion," Albert added. "That's the period about 500 million or so years ago, when life went quickly from simple-celled creatures to the kinds of animals we see around us today."

"Why is that a problem?" I asked.

"If an evolutionist says that the theory works its magic slowly, with tiny incremental changes over eons of time, then a creationist will hold up his hand, smile, and say, '*Cambrian Explosion*.'"

"I see," I said. "Meaning that those many life forms appeared much too quickly to fit the Darwinian view?"

"Right," Al said. "Of course, evolution needn't happen smoothly at all times. It's just unfortunate that we don't have a lot of proof available. Unless you count Lenski, of course."

"Who?"

"Richard Lenski. A brilliant man and what he set out to do was, frankly, audacious. He set up a rather interesting experiment. In fact, let me show you. You're about to see some very interesting glassware."

"What, like Tiffany or Baccarat?"

"Perhaps not so exquisite," Albert said. "Interesting. nonetheless."

The mist cleared, and I found myself looking down at a lab bench. On it, were arranged twelve simple-looking glass flasks.

"It's 1988," Albert said. "Richard Lenski has just set up these flasks with twelve populations of Escherichia coli. They started out as identical samples.

"For years, he watched the twelve communities grow, keeping a count of the generations, and watching to see what would happen as the bacteria replicated. The expectation was that, if Darwin was correct, there would eventually be differences in the flasks."

"How long did Lenski do this for?"

"He started in 1988. He might still be running the experiment, and I haven't checked for a while."

"What happened?"

Albert smiled. "There were developments in the bacteria. Sometimes, all twelve flasks demonstrated a similar change, and at other times only one or two did. The big news came around generation 31,500, when the bacteria in one flask were suddenly able to metabolize citrate—something that they'd never been able to do before."

"What had happened?" I asked.

"Lenski's team, led by Dr. Zachary Blount, asked the same question," Albert replied. "Their methodology included saving frozen samples of the contents every 500 generations, allowing them to be revived when needed. So, they went back and found that the attribute was identifiable somewhere around generation 20,000."

I frowned. "So, something changed back then, and was carried forward for many generations before becoming apparent?"

"That's certainly what it looks like," Albert said. "Needless to say, some scientists disagreed with the findings. And creationists had their own explanation."

"How could they possibly object?" I asked. "The science seems rock solid."

Albert smiled. "They claim that God must have given the bacteria the ability to metabolize citrate at the very beginning. They liken it to having a switch that lay dormant for a long time. The genetic makeup of that bacterium had the mechanism all along, and it just took specific environmental circumstances for it to switch on."

"Darned cunning, these creationists," I said.

MENDEL'S PEAS

My scientific studies have afforded me great gratification, and I am convinced that it will not be long before the whole world acknowledges the results of my work.

—Gregor Mendel

AS YOU'VE SEEN, IAN, THERE were many who were debating whether natural selection is correct, right?"

"Yes, Al, even nowadays."

He nodded. "In addition to that, however, there were those trying to find out how the process worked. They wanted to know what went on in an animal that would create the variability. Not just that, but could it be manipulated?"

I smiled. "I can tell that you have somebody specific in mind."

"I do. A monk, to be exact. In the Moravian city of Brno, there had been a fine monastery, and the town had kind of grown up all around it. The emperor, however, desired that prime real estate, and so expelled the poor monks to a site some distance away."

"An abuse of power," I said. "How unusual."

Albert nodded in agreement. "There was one lucky feature of the new property—it had a rather nice garden. That, in turn, suited the talents of a young man who arrived, with the intention of becoming a priest.

"In time that man, Gregor Mendel, was ordained even though he didn't seem all that interested in religious issues. He did have a curious mind, however, and was good with his hands.

"Around 1850, he had started quietly keeping a small community of mice, to see if he could produce hybrids. The supervising abbot found out what Mendel was doing and ordered him to stop. The other monks didn't like the idea of a monastery full of escaped rodents."

"Gregor worked with peas, didn't he?" I asked.

"He liked the fact there are many varieties of them. Around 1854, he was allowed to use the monastery's two-acre garden for his experiments. I'll twirl up some mist, and we can take a look.

As the air cleared, and we looked down at the garden, Al continued. "He visited nearby farms, and collected several dozen types of peas, then worked with them until he had what he considered to be pure strains. By which he meant that tall plants would only produce tall offspring, and so on."

"Then what? What did he discover?"

"He created hybrids, crossing tall with short, or mixing colors. That went on for eight years, and he kept meticulous notes along the way. In particular, he got the bright idea of crossing hybrids with hybrids, and that resulted in a surprise.

"For example, in situation where he thought he had eliminated tallness completely, it would pop up again. He was stunned. Where had that feature gone, and how did it reappear?"

"What was his solution?"

Al nodded. "He developed theories, later called *Mendel's Laws of Inheritance*, that organisms had both dominant and recessive features, and that an entire package of attributes was handed down from one generation to the next.

"He wrote it all down, explained in his careful and thorough way, in research papers that Mendel then read out to a small nondescript group made up mostly of local farmers. As his paper was circulated, everyone assumed that Mendel had simply confirmed what everyone had always thought about the way inheritance worked."

"Then what?"

"He tried to get attention for his work by sending it to leading figures in the field, but nobody paid attention to the seemingly simple religious man. For about thirty years, his papers lay ignored. Then the story got a bit convoluted. A few researchers, de Vries, Correns, and von Seysenegg, published similar results, claiming never to have seen Mendel's work."

"That could be true, if it was ignored for so long," I said.

Albert nodded. "Or it could have been plagiarism. It took until 1900, for a certain William Bateson to rediscover Mendel's paper and begin a movement to credit the priest with his important discoveries. Mendel had been quoted as saying that his work would be praised by the entire world, but that only happened long after his passing."

"Sad," I mumbled.

I continued to look down at the scene below us, as the priest slowly moved along the rows of plants, all carefully laid out in separate beds. He seemed to be counting and inspecting the plants, their heights and other features, and would occasionally pause to jot something down a small notebook.

Albert looked at me. "You of course know that this packet of genetic information is something that we call DNA."

"Right," I said, "Watson and Crick and their double helix. What can you tell me about that?"

"You'll be surprised."

DOUBLE HELIX

The nitrogen in our DNA, the calcium in our teeth, the iron in our blood, the carbon in our apple pies were made in the interiors of collapsing stars. We are made of starstuff.

—Carl Sagan

THIS DNA STUFF IS TRICKY," I said to Al. "How on earth did Crick and Watson figure it out?"

Albert smiled. "You're right. For example, suppose we had some kind of 3-D latticework, made of fine metal strands. And suppose I took a photo of it and showed it to you. What would you see?"

"Everything would be collapsed onto a two-dimensional image," I said. "I'd have a tough time figuring out what connected to what."

"Exactly," Albert said. "Now imagine that the photo I showed you wasn't even in focus. What then?"

"Even more difficult to see what's going on," I said. "Does this relate somehow to DNA?"

"It does," Albert said. "Let's go back and look at a key moment." He twiddled his fingers. When the mist cleared, I was looking at buildings in a city center, where the rain was gently drizzling down.

"I know where this is," I said. "It's London. I lived here for years. I think that's King's College."

"Correct," Albert said.

"I don't understand. Wasn't DNA discovered in Cambridge, and not in London?"

Albert nodded. "You have to understand that people the world over were trying to figure out the structure of DNA. For example, Linus Pauling at Caltech did important work, and revealed his idea of a single helix.

"Crick and Watson were shaken by that. They had believed they were in the lead. They redoubled their efforts to figure out the structure but were hampered by the poor images of the material."

"What kind of pictures could researchers get?" I asked.

"This story revolves around human personalities," Albert said. "You see, the Physics department at Kings was run by a man named Randall. He had brought in a certain Maurice Wilkins as his assistant, and together they hired a team of researchers. Randall, unfortunately, behaved like an emperor, making decisions without always telling the people who needed to know, and that had consequences. You'll see."

By now, Albert and I were hovering in a laboratory of sorts, looking down at a young woman.

"Wait," I said, "who's this? Crick and Watson are men."

"This is the tragic figure in the back story," Albert said. "She is Rosalind Franklin, a DNA researcher whom Randall recruited in 1950. He went ahead and hired her while Maurice Wilkins was on vacation."

"Wait," I said, "Wilkins was in charge of the work on imaging DNA? Then Randall gave the project to this new Franklin woman? Seems a little unfair."

Albert smiled. "It also seems that Randall told her that she'd be working under her own authority, but he neglected to inform Wilkins of that. What made it worse, perhaps, is that Franklin didn't endear herself to her new colleagues. She was argumentative and ill-tempered, and so many avoided dealing with her.

"Nevertheless, she took the idea of DNA imaging further, getting sharper and sharper images. You see, it had been realized that you could change molecules into crystallized form, freeze the atoms in place, then obtain X-Ray images of them."

"Two-dimensional ones only, of course," I suggested. "That was why they still had trouble figuring out how it worked?"

"Indeed," Albert said. "Up in Cambridge, Crick and Watson were using models to figure out the structure. They made physical structures using wire and colored objects, looking a bit like toys. Franklin thought that was childish and demeaning, so she worked in her own way, recording passages in her notebooks with comments about a big helix with several chains."

"Sounds like she was almost there," I said.

"She was. Now let's watch what happens next."

I looked down, and saw Franklin inspect a printed image closely for quite some time, then open a drawer, and carefully lay the document inside. She stood up, put on her coat and left the lab.

"I don't get it," I said, turning to Albert. "What's exciting about this?"

"Franklin was just handling the famous photograph fifty-one," Albert said, looking sadly at me. "We can leave here now, because I always feel sad thinking about what happens next."

"Why?"

Albert sighed. "Shortly after the episode we just viewed, Watson will arrive. Wilkins, not understanding the significance of photo fifty-one, will casually show it to him.

"But even if Wilkins hadn't made the mental leap, Watson knew exactly what he was looking at. In the train going back to Cambridge, he jotted down a sketch of what he'd seen on the photo. He had also immediately understood that DNA must be in two interlocked chains. Funny, really, that it took them all so long to figure that out. After all, so much in life is configured in pairs, isn't it? Male and female, yin and yang."

"I see," I said. "Watson had the answer, then."

"There was still much to do," Albert said. "It wasn't until 1953 that they had their complete model. The breakthrough came with the realization of how the backbone lay."

"I've read about this," I said. "Two strands, made up of sugars and nucleotides, right?

Albert smiled. "Sort of. At first, they had assumed the backbone ran down the center of the structure. It just seemed natural to them. After all, humans have one backbone, with stuff hanging from it, right?"

"Yes," I said. "But there were two chains. So, two backbones?"

"Indeed," Albert said. "The breakthrough was to realize that the backbones were on the *outside*, with nucleotides joining them together, like the rungs on a ladder. In realizing this, Crick and Watson made the second part of their major discovery."

"Meaning?"

"Meaning that they had also deduced how DNA replicated. It tore down the center, like a zipper pulling apart. Each separate strand is a complement of the other, so then attracts more material to recreate the original structure that had broken apart."

"Clever," I said.

"Clever enough for a Nobel," Albert said. "Watson and Crick got theirs in 1962. As did Wilkins, as it happens."

"What about Franklin?"

Albert frowned. "No Nobel for her. Wilkins later realized he should not have shown Watson her photo. He excused himself by saying that Franklin had already decided to leave Kings for another college. So, he rationalized that the photo wasn't hers to control in the first place. In 2000, Kings opened a new building named for Franklin and Wilkins. That's about all the recognition she got."

"Funny how those prizes can be influenced by chance events," I said.

"As was mine," Einstein said. "I didn't win it for relativity, but for the photoelectric effect."

"Why not for relativity?" I asked.

"Internal politics at the Nobel organization," Albert said. "Don't ever let anyone tell you that science and politics are different fields."

"Duly noted," I said.

Albert frowned. "It's sad really. Franklin died young, and you'll remember that a Nobel can't be awarded after a person's death. She had done incredible work but was never recognized for it.

"Of course, she wasn't the only one to lose out in this field. It was the Canadian, Oswald Avery, who was the first to figure out that genetic information was in DNA. Previously, folks had thought that it was in the cell's protein."

"Clever chap," I said. "Don't tell me he didn't get a Nobel."

Albert sighed. "He didn't. Many call him the most worthy scientist never to get the prize. He was nominated several times, but only ended up having a crater on the moon named after him."

"Well," I said, "being up in the stars. That's something."

"For God's sake, Ian, the moon isn't a star."

"You sound cross, Al. You know, I think you and I need a little break."

ISN'T LIFE WONDERFUL

The life of man is of no greater importance to the universe than that of an oyster.

—David Hume

ALBERT AND I WERE TAKING OUR BREAK from the science stuff. We were in Lambertville, New Jersey, enjoying brunch at a waterside café, and watching in fascination as a paddle steamer out on the river was going around in circles.

"What on earth is the pilot doing?" I mumbled. "Sailing aimlessly around like that?"

Albert chuckled. "He might be lost. Maybe he's looking for the exit to the Turnpike."

I laughed. "Well, at least he's enjoying himself. The river is lovely, the sky is blue. It's a perfect day."

Albert sighed. "It is interesting, isn't it, how we can find things to be utterly beautiful. The grass, the trees, and that favorite of the Creationists—the human eye, so fabulous that it can't have just evolved. The idiots."

"I know what you mean," I replied. "I had a most interesting dream about life and beauty last night. About a beautiful woman, to be exact."

Albert shivered. "I don't think I need to hear what goes on in your deviant slumber."

"It wasn't a graphic dream like that," I said. "Let me set the scene for you. I was vaguely aware of being in a lounge bar on the top floor of some fancy hotel, somewhere in a big city like New York or Hong Kong.

"I was quietly enjoying my Martini and indulging in a little people-watching, when my eye was caught by two gorgeous ladies who were obviously supermodels. They were standing at the bar, daintily sipping at their glasses of champagne—that kind with the tiny bubbles. One of the women was dressed in red, the other in blue.

"The one in red turned to look in my direction. Good grief, I thought, why can't these supermodels ever leave me in peace to enjoy my cocktail? Why do they always have to keep lusting after me?"

Al smiled. "Have I ever told you how deluded you are? First, you were going to write a brilliant book, and now you're consorting with models?"

I looked patiently at him. "This dream gets interesting, you'll see. I watched as the lady in red began to move elegantly in her high designer heels, and it was quite clear that she was walking towards me. She was also smiling my way, her lovely face just beaming with pleasure at seeing the magnificent specimen of maleness that I represent.

"Being of an analytical frame of mind, however, I immediately knew what I had to do before she reached my table—I had to calculate how many separate objects this lady in red comprised."

Al laughed. "So, there is a scientific component to this, after all?"

I nodded. "I started to figure it all out. There was her body itself, and it was a very nice one indeed. So, I counted one for that. Although, I did wonder whether she might have some kind of implant.

"She's too young for a pacemaker, I'd say, but she could have a metal support in an arm or leg, if she'd had a skiing or horse-riding accident of some kind. That kind of thing would set off metal detectors at airports, but then supermodels don't fly commercial, do they?"

"I do love how your mind goes off on tangents," Al said.

I continued. "Let's talk about her head. There was a hair clip covered in pearls. Hmm, was I to count the pearls separately? No, I decided not to. So, the body and the clip took me to a count of two. Same rationale for her pearl necklace. Hmm, maybe Mikimoto pearls, from that nice little store in the Ginza. The count was now three. Her two earrings made it five.

"She was probably wearing underwear—a bra, and possibly panties. I decided to verify that later, back at my place. The count was now up to seven. Hmm, she might have been wearing body jewelry, but for the moment I assumed that she was a nice girl and did without.

"A ring on her right hand. The two designer shoes, but no stockings. A Longines wristwatch. A small Fendi shoulder bag, the kind you could maybe just squeeze a credit card into. Hmm, that's a point, I didn't really know what was in the bag, did I?"

Al smiled. "How about a small lady's revolver, to deal with the unwanted attentions of deviants?"

I ignored him "So, that's about it. I had my answer, but it had to be expressed as an approximate value or a range, since I was unsure of a couple of things that might have been hidden from view. I was thinking the answer was . . . around twelve. I thought I'd go over this with her when she sat down, and we'd both have a good chuckle, and agree how clever I am. She was almost at my table by that point."

"What happened then?" Al whispered.

"I was crushed, Al. She sailed past the table, and it was then clear that her gaze wasn't on me; it hadn't ever been on me. I turned around in time to hear her squeal with delight and throw her arms around the neck of a man who had just entered the lounge. He looked like a kind of cross between Cary Grant and Humphrey Bogart.

"The lady in red and this Cary Bogart character made their way past my table, and they walked over to join the other supermodel at the bar, the one in the blue dress. They all laughed and joked together for a while. I couldn't hear what they were saying but I suppose they were discussing which of their various penthouses they'd go to later for their disgusting threesome."

"Rejected again," Al said. "Maybe you should try hunting for your victims in less ambitious locales. Perhaps at the local biker bar?"

"I didn't really mind, Al. You see, although they may have gone on to enjoy a night of pagan debauchery, I awoke uplifted in the knowledge that I had discovered a scientific truth. Namely, that the number of discrete physical objects that a typical supermodel represents is around twelve."

I smiled triumphantly at Albert.

"Quite wrong," he said, his voice firm. "And not just a little bit wrong, but so colossally wrong that it takes one's breath away."

"What on earth do you mean by that?"

Al smiled. "Your count of twelve is quite inappropriate. Catastrophically inappropriate, I'd say. Just think about it. If you're going to be talking about the nature of matter and life, then you'll be discussing issues like molecules and atoms, and so on. Right?"

"I suppose," I mumbled.

"Then you should talk about this lady in those terms. Figure out how many cells the lady has. And then how many atoms and fundamental particles that implies."

"How can I find that out?" I asked.

"I have the numbers here, as it happens," he said, taking a small notebook from his jacket pocket. "The average human body contains—no, wait, you wanted the average supermodel, and that's in another column. Here it is: the average supermodel body, non-Asian that is, contains seventeen trillion cells."

"Was that trillion, with a T?" I said.

"Right," he replied. "And wait, there's more. Each human cell contains about ten to the fifteen atoms. That's a one, followed by fifteen zeroes."

I sat there, quite stunned. "Al, that means a human body has an uncountable number of atoms."

"Uncountable for you, perhaps, but not for a real scientist," said Al. "Moreover, each atom contains multiple fundamental particles."

I gazed at him. "How on earth did life become so complex?"

"Well, dear boy, let's look at that."

HOW DID LIFE BEGIN?

You only live once, but if you do it right, once is enough.

—Mae West

ALBERT, I'M TRYING TO PICTURE the scene. How on earth did anything go from being just chemical matter to something that looks like life?"

Al nodded. "They used to think quite differently—they believed that all living things were moved by something called the *vital force*. This belief goes by the name of *Vitalism*."

"A force like the battery in an Energizer Bunny?" I asked.

"Sort of. Let's take a look at a most important moment in this part of the story. Excuse me, while I mist a little."

Moments later, I was looking down at the city of Berlin.

"It's 1828, and there's a lab up ahead that I want us to visit," Al said. "Let's go inside and meet Friedrich Wöhler."

"Tell us again, Herr Professor," said a student, "what it is that has you so excited."

Friedrich Wöhler looked around his class and smiled.

"I was conducting experiments with ammonium cyanate," he said, "and I found that I had accidentally synthesized urea. I find that most interesting, don't you?"

He looked round the classroom, checking the eager faces of the young students there.

"Herr Professor," said a young man, "isn't that a component of urine? If so, I'm not sure why this should excite us. We all produce that substance without difficulty, several times a day in fact."

There was laughter in the room.

"You are correct," said a smiling Wöhler. "However, there is one most interesting issue, if you pardon the use of that word. Namely, that it was always considered that urea could only be produced within living beings."

"I see," said another student, "so does this mean that other components of animals could also be simple chemicals? Or that life itself might be largely chemical, without the need for any magical vital force?"

"Well said, young man," the Professor replied. "We will have to see what others find out in their research, won't we?"

Al looked at me. "The history books paint Wöhler as becoming a fervent champion of the idea that all life is chemical. That probably wasn't his belief. But his discovery, albeit accidental, did mark the moment when many began to doubt the idea that there was some magical spark that caused life to form."

"I see," I said. "So how did life really begin?"

"Back in 1953," Al said, "the Miller-Urey experiment, as it's known, involved a solution of water, hydrogen, ammonia, and methane."

"Why that mix?" I asked.

"This was considered to represent the environment in the early history of our planet. Several billion years ago, in fact."

"Then what?"

"They heated the mix, then introduced electric sparks, to simulate lightning strikes. Some believed that the early Earth had significant storm activity, you see. The result of their experiments showed that sugars and amino acids had formed. And those, as you know, are among the building blocks of life."

I smiled. "Well, that settles it, then."

"Not quite," said Al. "Since then, they've determined that the land-based environment of the early Earth was comparatively low in hydrogen."

I frowned. "So, what, we delete the idea from our list?"

"Not yet," Al said, chuckling. "Others say that volcanic clouds could have thrown up enough of the required substances to make those reactions happen after all."

"My head is spinning," I said. "Can't we get clarity or certainty on how life began?"

"Nope," Al said. "There are several other ideas, and they're also good ones. For example, some theorize that the substances we described were not kick-started into life by lightning, but were sheltered in mud or clay for long periods, where biological units were gradually assembled by chance.

"Then you have the folks who point to deep sea vents. Those are locations in the oceans where the substances we've been discussing are constantly oozing up through fissures in the sea bed."

"Why is that idea favored?"

Al nodded. "Because there are such places even today, and the material gushing out of them nourishes all sorts of sea life."

"I see."

"We're not done. There are notions of life beginning in the frozen areas of the planet, or even of biological materials coming into our atmosphere from beyond the Earth."

"Beyond the . . . what?" I said, astonished. "You're not going to start talking about Martians are you, Albert?"

"Why not?" he said, an evil smile on his lips.

CONTAMINATION AND WELLS

We have your satellite. If you want it back, send twenty billion in Martian money. No funny business or you will never see it again.

— Reportedly seen on a wall at NASA, after contact lost with the Mars Polar Lander in 1999.

"WHERE IS THIS, ALBERT?" I asked, as the mist cleared.

"The Columbia Broadcasting Building in Madison Avenue," he replied. "The date is October 30th, 1938, and the man at the microphone is Orson Welles."

"I've heard of him. Citizen Kane, right? A great movie, and I remember seeing it when I was a boy."

"I agree about that movie," Albert said. "This moment, however, is what really made Welles famous. He's about to narrate a version of H.G. Wells' story *The War of the Worlds*, and it's going out as an episode of a radio series called *The Mercury Theater on the Air.*"

I watched as the imposing figure of Welles stood at the microphone, beginning the broadcast in a most serious tone of voice.

"We know now that in the early years of the 20th century, this world was being watched closely by intelligences greater than man's and yet as mortal

as his own. We know now that as human beings busied themselves about their various concerns, they were scrutinized and studied, perhaps almost as narrowly as a man with a microscope might scrutinize the transient creatures that swarm and multiply in a drop of water."

Albert smiled. "In preparations for the episode, Welles and others thought that just reading out the H.G. Wells story would sound too dry. So, they decided to turn it into a journalistic-style news bulletin."

"Wait," I said, "Isn't this where the population of the USA went into a panic, thinking they were really being invaded by Martians?"

Al shrugged. "The idea of a huge panic has been exaggerated. But I recall that quite a few people were fooled. The show had begun with music from Tchaikovsky's First Piano Concerto, followed by Welles being introduced as the star of the broadcasts. Soon after he read out the section you just heard, they began interrupting with phony news bulletins about explosions on the planet Mars.

"But they changed the location of Wells' story from England to the USA, and that made it more believable for the radio audience. There was talk of a cylindrical object landing in New Jersey, with the military surrounding and attacking it. They described people falling like flies in Times Square, and large Martian machines striding across the Hudson River."

"I see," I said. "So, if you tuned in after the show's introduction, you could have thought the whole thing was real?"

"Right," Al said. "Let me fast-forward a bit."

The mist cleared again, and I looked down at an ordinary-looking room.

"Where's this?" I asked.

"One of the anterooms in the broadcasting center. The actor sitting there is Stefan Schnabel. He's finished his part of the performance and is relaxing but he's about to be interrupted. Just watch."

I kept looking, and saw some policemen arriving.

Al smiled. "They are about to ask the network to stop the broadcast, because too many people are contacting the police, asking them what's going on. One of the tricks used by Welles was to delay advertising breaks to make the show sound more real. The longer it went on, the more people became scared.

"You know, Wells and Welles met, when they both happened to be in the same area. I bet they had a good laugh at what Orson had done with the story."

"Interesting, Al, but why did you show me this?"

"Almost forty years later, NASA were sending Viking probes to the red planet and, perhaps because of that radio show, and other speculation about the canals on Mars, they felt they just had to test for life there.

"They used techniques that had initially been developed to assess the drinkability of municipal water supplies. On Mars in 1976, that equipment deposited some prepared material onto a small amount of the planet's soil, then looked for certain microorganisms in the gas that was released. When those results were examined, it did seem as if the basic building blocks of life had been identified."

"What do you mean, *did seem as if*?"

"NASA's senior staff preferred not to believe their own work," Albert said. "The media didn't want to touch this, since the news might have sent Christians who believed in a Creator into a total panic. Remember, most of those people still think that their God created life here on Earth. And *only* here, in this tiny corner of the universe."

"I see," I mumbled.

"More and more tests were devised. On meteors and lunar rock, for example. Scientists frequently discovered signs of life's components, but there was always one word they heard from their skeptics."

"Fabrication?" I suggested.

"Occasionally that word," Al said, "but more often it was *contamination*. The idea being that, no matter how careful scientists were, earthly contaminants just had to be creeping into their equipment or test materials."

"What did scientists say about that?"

"They worked harder and harder to eliminate any possibility of contamination. They developed techniques to extract samples from the center of meteoric materials, in absolutely sterile conditions. They went up above the Earth's atmosphere and took samples from space dust unaffected by the Earth. It made no difference."

"Nobody believed the results?"

Albert nodded. "That's right. But the totality of that research work gave rise to a new theory that they call *Panspermia*. It's the idea that what's been found in our solar system is by no means exceptional.

"Those who believe in Panspermia think that we are likely to find the building blocks of life just about everywhere in the cosmos. And this has a particular advantage when explaining the emergence of life here on Earth."

"How so?"

Albert smiled. "Creationists latch on to a particular issue that has been confounding Darwinists for a long time, namely that the early environment of our planet was positively hostile to the formation of life.

"As soon as biological components connected up, they'd have been destroyed. The skeptics of science say that life just can't have emerged here—it would have required too much time, therefore . . . God. That in turn obliged scientists to scramble around looking for exceptions, such as those vents in the sea floor where conditions were more favorable."

"I get it," I said. "With Panspermia, the long evolution can take place on some distant planet, where conditions are better, and then the material can be scattered throughout the cosmos in black hole emissions or supernova explosions."

I thought for a few seconds, then looked at Al. "But hang on a minute. How could life forms, even microscopic ones, survive in space?"

Al smiled. "In the 1950s, clusters of bacteria were discovered alive and well, even thriving, in very powerful radioactive waste that had never been present on earth before the twentieth century. This implies that, at some earlier point, those bacteria had gradually evolved protections for surviving radiation, such as that found in space.

"That's not all," Albert went on. "It's known that types of fungus can live through colossal bursts of radiation and can survive in a vacuum. There are also microscopic life forms that can live through extreme impact, such as that of crashing into the surface of the earth."

"I see," I said. "This means that the origins of life could have developed elsewhere in space, where they conditions were more favorable than our early planet's. They could then have arrived here on our much younger planet, already well on their way on the evolutionary ladder."

"Agreed," Albert said. "There is currently agreement that, although our planet is under five billion years old, the universe may be nearer fourteen billion."

"Plenty of time for Darwin's evolution to do its thing out there," I said, gazing up at the skies.

INEXACT COPIES

Men are generally more careful of the breeding of their horses and dogs than of their children.

—William Penn

ON SEVERAL OCCASIONS, I HAD INVITED Albert to my home, and from time to time he'd turn up unannounced. One day, I found him browsing through the volumes in my library and, when he heard me entering the room, he turned and frowned.

"This is what you're basing your ideas on? This motley crew of books?"

"You like them, then?" I said.

"You really don't understand sarcasm, do you?" he replied.

"Don't have much time for it," I said. "Why don't you sit on the terrace, and I'll fix you a drink."

He went outside, then settled himself on a chair and gazed out at the valley.

"Here," I said, carrying a tray with snacks and a couple of vodkatinis.

"I don't drink," Al said, looking sadly at me. "You do realize, I hope, that I'm a figment of your imagination?"

"That hadn't occurred to me, Al," I said, smiling. "I'll just have to drink the two tinis myself."

"What did you want to discuss today?" he asked.

I sat for a moment and collected my thoughts.

"I think," I said, "that we've made good progress. You and I have talked about how life could have come to be. We saw the work of Mendel, Darwin, and so on, and we've looked at the ideas behind DNA, and how the first signs of life could have emerged on our planet. I've heard it said that every living thing on our planet shares the mechanism of DNA."

"True," said Albert.

"Then, what was it that created the diversity?" I asked.

"Ah," Albert said, smiling at me. "You want to know about replication, don't you?"

"I do, Albert."

"Picture it, then," Albert said. "The DNA ladder structure collects material on its many rungs. The whole thing, although it seems infinitesimally small to us, gets to be quite heavy and cumbersome. So, at certain points along its length, it starts to separate."

"Why?"

"Because the various points of separation happen to be made of materials with weaker bonds, that's why. The ladder flexes, then pulls apart at certain predictable locations, like a long zipper."

"I get it," I said. "The two backbones are complementary images of each other, and the now-exposed split ends of the rungs will only attract a specific type of material. So the double ladder rebuilds, exactly as it was before, to split and replicate again and again."

"Hmm, not quite," Albert said, smiling gently. "Think about how hard it is to copy something."

"What do you mean?"

"The greatest artists in the world can still have the toughest time faking another person's work. Heck, they could even have a tough time producing a flawless copy of their own paintings. It's hard."

"What does this have to do with DNA, Albert?"

"Only that the replication of DNA is sometimes inaccurate. If it's wildly inaccurate, the life form in question fails to survive. Did you know, for example, that as many as fifty percent of a woman's fertilized ova are naturally aborted, most of the time before the woman is aware she's pregnant? And spontaneous abortion can occur after that too."

I smiled. "All because of imperfect copying processes."

Albert nodded. "You know, they say that all human life on this planet may have come from just a single source."

"Why just one?"

"It's true that there could have been several, but then competition would have reared its ugly head. The stronger forms would have survived, while the less capable ones fizzled out."

"Our friend Darwin again," I said.

"Let's recap," Albert said. "We saw how the mechanisms of life work, at least the broad lines of it. We've speculated on how life originated, or arrived here on earth."

I nodded. "Also how the replication errors produced all that we see around us. Birds and bees, and so on."

"There's another important thing to know about life," Albert said.

"That being?"

He looked out onto the valley and, for a moment, watched the red-tailed hawks circling above. "It's just that," he said, "there may be no such thing."

"What does that mean?"

He laughed. "It's that old idea that there's no distinction between life and non-living things."

"We talked about that the first time we met," I said. "What do you know about it?"

"Have you ever looked at a rock?" Albert asked.

"I usually prefer watching the football on TV," I said.

ON THE ROCKS

A fool can throw a stone in a pond that a hundred wise men cannot get out.

—Saul Bellow

CONSIDER THE LOWLY BOULDER," Albert said. "It lies in the river bed, with the water rushing over it. Would you say that it's alive?"

"Not really," I replied. "I mean, it doesn't move, does it? It shows no signs of life?"

"What about when there's a heavy storm, and flood waters surge down the valley? Does the rock move then?"

"Hmm, it might. Or if it were in a glacier, it would be carried forward. But ever so slowly, Albert."

"Suppose you could view all this from above?" he said. "And that you speed it up, so that thousands of years can be viewed in, say, one minute. What would you see?"

I thought about it for a moment or two. "I guess," I said, "that it might look as if the boulder is making a journey of its own volition. Moving slowly down the mountain, or along the river bed, perhaps even changing direction as it encounters obstacles."

"I agree," Albert said. "There's another issue. What would you see if you could peer into the tiny pieces of matter that make up a boulder?"

I looked at him. "Based on our earlier discussions, I imagine we'd see molecules of this or that, and that they contain atoms of certain elements."

He smiled. "And inside the atoms you have fundamental particles whizzing around in a furious dance, propelled by the energy that pervades space. So, not such a peaceful scene now, is it?"

"You're right, Albert," I said, thinking through what he'd said, "and the atoms in the boulder are no different from those in our bodies, are they?"

"Not if they're the same element," he said. "So, you could be forgiven for thinking that what we call living matter and inanimate matter are not so different."

"No difference, then, between life and non-life," I said.

"Not a new idea, I'm afraid," Al said. "Even way back, thinkers like Heraclitus had such ideas. He was an interesting man, and Diogenes reports Heraclitus as having said of himself that, 'as a boy he knew nothing, but later knew everything.'"

I laughed. "My kind of guy."

"Heraclitus believed that everything was in flux. Nothing stayed still, and all was constantly in motion. His famous saying was that you can never step in the same river twice.

"Some modern philosophers with a sense of humor say that he should have said you can never step in the same river *once*. Meaning that there is no such thing as a river. Nothing ever stays in one form long enough to deserve a name that separates it from anything else."

I smiled. "That would mean that our planet, and all that's in it, is just one big bowl of a ceaseless, churning activity."

"Precisely."

"That's pretty deep thinking, Albert," I said.

"It's what I do," he said, smiling gently.

"What you've said makes me think of Buddhists, Albert. Don't they say that we are all one?"

"Careful, Ian, don't get pulled back into religion."

"I've never thought of Buddhism as a religion, Albert." I said. "It's not totalitarian enough. Not enough holy wars, crusades, or divine commandments."

"Thank God for that," he said.

SETTLED SCIENCE

Believing in progress does not mean believing that any progress has yet been made.

—Franz Kafka

I MISSED ALBERT, BUT IF MY BRILLIANT BOOK was ever going to see the light of day, I had to keep plugging away. I spent several days organizing my thoughts and tapping them into my computer, then finally collapsed into bed.

I slept the sleep of the just then awoke to a fine and sunny morning. After a splendid, cardio-arresting English breakfast of bacon and eggs, I was doing the dishes when a knock came at the door. A rather heavy knock, I thought. Hmm, I wonder who that can be?

On opening the door, I found myself facing two men in dark leather coats. I must say, they looked a bit menacing.

"Can we come in?" said one of the men.

I thought it best not to get all pedantic and start one of those whimsical conversations about the difference between *can* and *may*. These men didn't look like they had a sense of humor. Besides, it didn't seem like they were really asking my permission.

They entered the house, and we all made our way into the living room.

Once we were all seated, I looked at them and asked, "How can I help you?"

"We're from the NSSA," said the second man, flashing a badge at me.

"I see," I said. "What would the National Security Agency want with me?"

"Not the NSA," said the first man. "The NSSA. We're from the National Settled Science Agency."

"The settled . . . what?" I said, confused.

"We're charged with ensuring the stability of scientific ideas in our country," said the second man. "Our agency was originally set up to protect the concept of global warming, which was being attacked by all those deniers and flat-earthers. We're here now because we've noticed that you are starting to construct some disturbing concepts of your own, and we're worried that they might disrupt the conventional wisdom."

"How would you know what I—"

"In your computer," interrupted the first man, as he pointed to my laptop on the table. "Those ideas you've been keying into your laptop."

"But I'm the only one with access to it," I said. "Everything is password-protected."

The two men looked at each other and smiled quietly.

The second man then gazed sternly at me. "Why don't you explain yourself. Tell us what these ramblings of yours mean."

"Er, okay," I mumbled, wondering where to begin. "Well, the first major concept I keyed in was this: *If there is a creator, then nothing requires explanation.*"

"Can you expand on that?" said the first man.

"Of course," I said. "It means that . . . if there is a God, then everything in the universe is the way it is because He wanted it that way. The sky is blue, and not green, because God wanted a blue sky.

In a similar way, horses have four legs and not five, because He likes the number four.

"So, if one believes that idea, then there is no need to seek explanations for everything around you." I paused, then added, "This, you see, is the situation that most people find themselves in, what with the majority of the human race belonging to one or other of the creator-centric faiths."

"Understood," said the second man. "What's the next idea on your list?"

"It's a corollary of the first statement," I said.

"Isn't that a heart attack or something?" said the first man.

"No," I said, hesitating for a second, "that would be a coronary. A corollary is an idea that flows from a previously stated idea or theorem. So, my second statement, flowing from the first, is this: *If there is no creator, then everything requires explanation*. What that means is that those who don't believe in a god have work to do. They have to explain to themselves why the cosmos is the way it is. In fact, they have to explain why things even exist at all."

"Why do they have to do that?" said the second man. "Why can't they just be good law-abiding citizens, and keep their thoughts to themselves?"

I was worried now. "Are you suggesting I'm breaking the law in thinking what I'm thinking?" I said. "If so, please tell me which law."

"That's classified," said the second man. "Besides, didn't you know? Every single person in the United States commits two or three felonies every day."

"That's the beauty of our legal system," said the first man, with a sinister smile. "The laws are so numerous and complex that nobody really understands them." He paused, then added, "It makes our job real easy."

Hmm, I thought, another grammatical error, but let's soldier on. I tried to regain my self-control.

"The next idea, please," said the second man.

"Yes, of course," I said. "It's this: If there is no creator, then nothing can ever have been created."

"That doesn't make sense," said the first man. "How do you explain the rivers and trees, and the birds and bees? Seems to me that lots of stuff has been created all around us."

"What I mean by that," I said, now becoming more nervous, "is that nothing can ever have been designed or constructed to start the cosmos. Nothing with any complexity or structure whatsoever. Because . . . there's no creator. The beginning of the cosmos, then, had to be something extremely simple.

"And that brings me to the next idea on my list: If nothing was ever designed or constructed, then the universe probably came from . . . nothing."

"I don't like the sound of that," said the second man, frowning. "I'm starting to suspect we might have a serious violation on our hands here." He turned to his colleague. "What do you think?"

But before that man could reply, I had a sudden inspiration.

"Look," I said, "you told me you were in charge of protecting settled science, right?"

"That's correct," said the first man.

"Well," I said, "the idea that the universe came from nothing has recently been moving into the settled science arena. It didn't use to be, I grant you, but it is now. Many physicists favor the idea."

"Wait a minute," said the first man, "you've overlooked something. If the universe came out of nothing, then what was there before all this?"

"Ah," I said, "that brings us to the biggest problem of all, namely *time*, and I haven't started figuring that out yet." I smiled at them. "Maybe Al can help me with it."

"Who's Al?" said the second man.

"Albert Einstein," I replied. "He is my guide as I've worked on all these issues. He's been taking me back in time, to see great scientists making their discoveries. It's been wonderful, working with him."

"Albert . . . Einstein?" said the first man, looking sideways at his colleague. "Time travel?"

The two men looked at each other, then rose from their seats and began edging towards the door.

"Well, okay then," said the first man. "I think we've finished here for now." He reached the door. "Don't bother to see us out."

Before following his colleague, the second man turned towards me. "We'll be sending some other Federal employees to, er, help you. You'll recognize them, because they'll arrive in a white van, and they'll also be dressed in white."

The first man added, "And they might ask you to wear a strange-looking jacket with funny straps. But don't worry, it'll all be for your own good."

"One more thing," said the second man, "on no account are you to leave this building."

THE ILLUSION OF TIME

Does time exist?

TICK TOCK

What is time? I know what it is if nobody asks me.
But if anyone does, then I can't explain it.

—St. Augustine

OF COURSE I LEFT THE BUILDING. I mean, I m not stupid, am I? A van and people in white uniforms, carrying a straight-jacket? I knew immediately which government agency the two men were describing.

It was the IRS.

I pushed those horrific thoughts aside. An important moment had arrived. It was finally time to think about time itself, that most elusive of concepts. If I can puzzle out what time really is, I thought, I will have done the world a great service.

Not to mention ensuring myself a knighthood Yes, I could look forward to the day when the Queen herself would dab my shoulder with her sword and say, 'Arise, Sir Ian, it was about time that someone figured out time for us all. Do stay for tea and cucumber sandwiches. If you have time, that is.'

After my inevitable knighting at the Palace, getting dinner reservations at top restaurants will no longer be a problem.

"It's Sir Ian calling again. Table for two please. Me and just one supermodel this evening."

Suddenly, Albert caught up with me again. I wasn't sure how he knew what I was doing, or when I needed him. He just seemed to show up when required. Maybe he was a genius after all.

"There you are, my boy," he said, smiling at me. "I was reading the notes in your computer. Seems you're ready to think about time?"

I was starting to get really annoyed that everybody was able to peer into my laptop and read my stuff. But I decided to go along.

"Yes," I said. "Assuming you have time for time?"

"I have nothing but," Al said, chuckling. "Let me take you to a little store just down the road. Maybe the owner can help you."

We left the house and began strolling in the nearby town.

"What, no mist?" I asked.

"No, this is in the present," Al said.

If Al hadn't been with me, I would probably have missed the little clock shop. It was tucked away in an alley and had an ornate sign on the door, reading, *Tick Tock Clocks. Time is on Your Side.*

I went down the alley, found the door, and went in.

"Come in, come in," said a jovial voice. "I have plenty of time for you."

I walked further into the store then found myself looking at a rather kindly old man.

He smiled at me. "Get it? Plenty of time?" He waved at all the clocks that were displayed around the store.

"I saw your store's name out front," I said. "Time is on Your Side? You must be a Rolling Stones fan then? They had an old song, didn't they? *Time is on My Side?*"

"Indeed I am a fan," said the man, who I was now figuring to be in his eighties. "I've supplied the stars with clocks for years. Lots of celebs on my customer list. Very good for business, you know. And I do indeed have a nice little thing going here. Now, how many clocks do you want today?"

"I dunno," I said.

"Tell you what," he said, "I'll make you a deal you can't refuse. Buy one clock, and I'll let you have a second one for the same low price as the first." He chuckled.

"I'm not sure I need two clocks," I said, smiling quietly. "You know what sailors say, don't you? Never go to sea with more than one compass, because then you'd never know which one is correct. The same would be true of clocks, wouldn't it?"

"Ah, sailors and clocks," he said with a grin. "Yes, sailors of old loved their clocks, didn't they?" He paused, then added, "They used all sorts of tools to navigate in early times, before they had accurate clocks."

"What?" I said. "I was only joking about sailors."

He laughed. "No, no. Navigation was done using all sorts of tools and methods, and none of them was very accurate until they had good clocks. My own favorite story on early navigation is of the shipwreck at Antikythera in the Mediterranean. You've heard about that?"

"No," I said.

He nodded. "In 1900, fishermen were sponge-fishing off the coast there. One of the divers found an ancient shipwreck that was full of statues and pottery. What puzzled the men, though, was finding a metal device, about the size of a shoebox. It had all sorts of gears and wheels."

"A computer?" I asked.

"That's what the crew wondered at first. They thought they'd found some kind of calculation device. It looked so modern and sophisticated that they figured it might have fallen off a more recent vessel and landed by chance on the much older shipwreck below.

"That turned out to be incorrect. It really was an old computer of sorts, with inscriptions in Greek dating to the first century. Sailors back then had used the machine to track the stars, and to navigate.

"That method alone still wasn't totally effective, though. They had to wait many centuries for the first really accurate chronometers to appear, to let them find their way with certainty."

"I don't understand," I said. "What do clocks have to do with navigation?"

"Good question," he said. "You see, sailors could figure out their latitude quite well, just by watching the stars and using devices like the one in the shipwreck. But to find their longitude, they had to use clocks, to figure out how much time had elapsed from their previous checkpoint." He chuckled. "There are no mile markers out on the ocean waves, you know."

I nodded. "I see. Early clocks weren't good enough?"

"Not by a long way. In the past, a typical clock might be off by many minutes in a day, and that could put a vessel off by a hundred miles or more.

"Then, in the early eighteenth century, a clockmaker called John Harrison developed much more accurate chronometers. They say he went along on an early test voyage, and the crew were skeptical when he told them that the destination would appear on the horizon the next morning. But it did, and all thanks to his clock."

"Fascinating," I said. "It really is. Thanks for telling me about that."

The old man started walking round the store, pointing out clocks to me.

"Now, which of these beauties takes your fancy today?" he said. "I have grandfather clocks against the wall there. And a couple of smaller, grandmother clocks. Family discount, if you buy a pair.

"I have carriage clocks, cuckoo clocks from Austria, and some from Switzerland. There are mantel clocks, and I even have a clock made of wood in a kit that you can assemble yourself."

"All very interesting," I said, "but in coming here I was hoping to find someone who could talk to me about time itself. What is time, exactly?"

He chuckled. "Where to begin? It was Eddington, wasn't it, who said in 1927 that time was a force like an arrow, carrying everything along with it? Many agree with him and think that time flows in only one direction.

"Of course, you do get the lunatics who say that the universe is really a stack of different universes, and that one could flip from one to the other and go back in time that way. Drivel, of course. The universe is much, much simpler than that."

"I see," I said, getting more excited now. It looked like this man might be about to divulge important information.

"Then there was Einstein's work," he said. "He theorized that time and space were somehow connected, in a *space-time continuum*. A ridiculous phrase, that."

I looked around the store, but Al was nowhere to be seen.

The old man continued. "Einstein also said that time wasn't the same for all players in the cosmos, and that anyone traveling faster would age less quickly. In other words, time would pass more slowly for them."

He chuckled. "He was eventually proved correct about that, in 1971. They developed some really accurate atomic clocks, synchronized two of them, kept one on the ground, then sent the other round the globe on a plane. After that journey, when the two clocks were compared, the clock from the plane was a tiny bit behind its partner. Time had indeed moved more slowly for it."

"Brilliant," I said, "but that raises the same old question in my mind—if time can vary like that, then what the heck is it?" I paused, then added, "Is it even possible to know the answer?"

"Hmm," the old man said, looking serious now. "It is possible. It is. But you have to be mentally ready to understand."

"What?" I said, surprised. "Do you mean that you know what time is, but that you can't tell me?"

"I know it sounds unfair," he said, "but if I give you the answer, you might not understand it. Even if you do, you still might not believe it. For this issue, it's always best for the thinker to arrive at the truth by finding their own path." He paused, then said, "But I do have a suggestion for you."

"Yes?" I said, now eager to learn more.

"I will give you a small sundial, free of charge. Except for shipping and handling, of course, which will be fifty dollars."

"Shipping and handling?" I said. "But I'm standing right here."

"Of course, of course," he said, chuckling. "My little mistake. The shipping is free." He smiled. "It's the handling that costs fifty dollars."

I grudgingly went along with his scam. After all, his knowledge might come in handy. I paid him the fifty dollars.

"I'd like you to set it up outside your home," he said, "and then watch it to see how it operates."

"I don't understand," I said. "What would I learn by watching it? A sundial has no moving parts."

"It does, actually," he said, with a smile. "You will find that out, if you look for long enough. It's also a form of meditation that might help you find inspiration."

I thanked him for his time and his kindness, and left the store with the small, metal, overpriced sundial.

Once back home, I set it up outside, in the garden.

I placed it just by a window, so that I could look down on it as I worked. Who knows, maybe the shopkeeper was on to something. He seemed so sure of what he was saying.

MOVING PARTS

Even a stopped clock is right twice a day.

—Marie von Ebner-Eschenbach

ALBERT HAD FOUND ME AGAIN. Darn, he is a cunning old thing, because he had arrived just in time for lunch.

I placed quiche and coffees on the table by the window in my kitchen, and we sat there, looking out at the sundial.

"I still can't figure out what the old clock man was on about," I said. "He said that thing out there has moving parts, but I just don't get that."

"Why don't you keep watching for a while," Al said, a smile on his face.

I did as he said, keeping an eye on the shadow on the clock face. After half an hour, I noticed the change in its position.

"Well, it's now at twelve-thirty," I said.

"You don't see the moving parts yet?"

"No," I answered. "I told you already—there aren't any."

"You're particularly dense today," Al said. "Let me end your torment. A sundial has one and only one moving part, and it is our planet, the Earth."

I looked at him, then back out at the clock, then laughed.

"I see what you mean now. The sun shines down on Earth and, as the planet revolves, the shadow strikes a different part of the clock's surface. Clever."

"Not really clever. Elementary."

I thought some more. "So, Al, what does this tell us about the passing of time?"

Al looked at me. "If I told you that I would meet a train at the station at seven, what does that really mean?"

I looked back at him. "Doesn't it mean that when enough time has passed, you and the train are at the same place? Or maybe I don't fully understand your question."

"It may simply mean," Albert replied, "that the movement of the energy powering my watch pushes the hands to the seven o' clock position. In a similar way, the energy does the same thing for the train driver's watch. And the both of us keep an eye on that, and we adjust our own movements to arrive together on the platform when the hands are at seven."

"Okay, but what does that mean, Al?"

"It means that the important thing is not a river of time as an independent force. Rather, it's the underlying energy of the cosmos powering our machines, including clocks, that create our sense of this thing called time."

I looked at him, not quite understanding where he was going with this.

"I'm afraid the world became quite confused," he said, "when my work started people talking about a space-time continuum."

"I never did understand that phrase," I said.

"It was to help people with the relativity of the two supposed forces. If you move rapidly through space, you move slowly through time, and the converse of course."

"Yes, I remember this," I said. "They proved you were right about that, didn't they? The old clock man described how they tested the two atomic clocks, after one of them had been flown round the globe."

"Exactly," Al said. "I do wish I had been much more specific about what time is. It would have helped an awful lot of people."

I considered what he'd said for a minute or two.

"Okay, Al," I said, "but everyone seems to agree that there is a force called time. After all, we get old and die, don't we? Surely, they'd say, that's because of an underlying current of some kind, carrying everything along its path?"

"Perhaps," Al replied. "Of course, there is that thing called entropy."

"I've never fully understood that?"

"Well, Ian, we will need to talk about ping-pong balls again, I'm afraid."

I frowned. "Ping-pong balls?"

"Yes," he said, chuckling. "You found them useful earlier, didn't you? So do most physicists."

FALLING TO BITS

Only entropy comes easy.

—Anton Chekhov

LET'S IMAGINE," SAID ALBERT, "that we make a two-foot cubic box out of something like chicken wire. You know, a wire mesh, but with really small holes so that a ping-pong ball cannot pass through."

"Okay, Al, what then?"

"Then we build another one that is exactly the same," he replied.

"Got it," I said. "Two cubic boxes of wire mesh. Now what?"

"We make a hole in the top of each box, say about four inches in diameter. Then we take one hundred ping-pong balls, individually numbered, and we pour fifty into one box, and the rest into the second."

"I'm picturing it," I said.

"We tip the boxes onto their sides and push them together so that the two holes are joined up. Then we place electric fans all around the perimeter and turn them on. What do you think will happen?"

"The fans are powerful?" I asked.

"Very," Al said. "Enough to send the balls bouncing around like crazy."

"In that case," I said, "they would bounce off the walls of wire mesh, sometimes colliding with each other. But I think I see what you're after, Al. Some of them would also pass through the combined openings, wouldn't they, and go into the other box? And that could happen in both directions."

"Right," he said, smiling. "If we let the fans run for a while then switch them off, what do you think we'd have when the balls settled down?"

"We'd get a slightly different distribution," I replied. "It might be still around fifty-fifty, but some of the balls could have moved into a different box from where they started."

Al chuckled. "Could we have one hundred balls in one wire box, and none in another?"

"No, Al, most unlikely. Not unless the fans are greatly imbalanced."

"Assuming the fans are blowing evenly," Al said, "and assuming also that we repeat this experiment millions, even billions of times, is it impossible that we'd end up with one hundred in one wire box?"

I frowned. "Not impossible, I'd have to say."

"In fact," Al said, "there is only one way for one hundred balls to end up in one of the two boxes. But how many ways are there for a box to contain just one ball?"

"Er, one hundred ways, right?" I said. "Any one of the hundred balls could be the one that ends up on its own."

"And how many ways to have two balls in a box?"

"The first ball, Al, would be any one of the hundred. And the second would be any one of the remaining ninety-nine. So, nine thousand and nine hundred combinations."

"Very good," he said, nodding. "Maybe you're starting to get my point. There are very few ways to have something complex appear—like the hundred balls in one box—and many more ways to have smooth distributions of something. Like energy, for example. Or energy that combines to produce life forms, like people.

"Not just that, but if you do reach a point where you have one hundred balls in one box, then if you keep running the experiment from there, the balls will eventually degrade back to a more normal distribution."

"The second law of thermodynamics," I said, smiling at Albert.

"Which says that in a closed system, entropy will inevitably increase over time."

I nodded. "Entropy being the state of disorganization?"

"Roughly speaking, yes," Al replied.

I nodded. "If that's correct, then how would life ever have appeared here on Earth? We're a closed system, so energy should just have become degraded. Instead of which, we have birds, bushes, fish and trees."

"We are not a closed system here on Earth," Al said. "We receive a massive flow of energy from that orange thing in the sky, the sun."

"I see," I said.

"Not just that," Al said, "but the Second Law does not forbid isolated arrangements of complexity from forming here and there. It's just that, over time, energy will tend towards simplicity."

I smiled. "That is why things live and die? Our energy has managed to coalesce into something complex, but it will degrade over time?"

Al nodded. "Even if humankind doesn't blow itself up, the sun will eventually burn out, and our human energy will dissipate as we all die. That will happen throughout the cosmos, with suns collapsing and dying."

"The cosmos will end darkly and quietly, then," I said sadly.

"Not with a bang but—as they say—with a whimper," Al said.

We were quiet for some time, then Al looked at me. "Something stronger than coffee?"

"I was thinking the same thing, Al."

A POEM

The sum of all sums is eternity.

—Lucretius

I WAS STANDING AT THE EDGE OF THE WATER at Sandy Hook in New Jersey. It's a place I used to go to years ago, whenever I needed to think. There's something calming about watching the waters of the Atlantic, especially in the cold of winter.

I felt that someone was approaching, and I turned to see Albert, looking his usual scruffy self and smiling at me.

"I haven't seen you for weeks," I said. "I thought you'd gone forever."

"I appear when I'm needed," he said. "I've been watching you all this while, and I sensed that you were stuck. You know, in need of a gentle nudge to move you forward."

"If you push me now, I'll fall into the sea," I said, laughing. "But you're right, Al, I've reached an important point in what I'm doing. I've realized that science can't help me anymore. It's something else that I need now."

"I see," Al said. He looked at me, his expression more serious now. "You're not the first to be in this predicament, and there is something we could try, if you're willing. I'd like to take you back to observe a very interesting man—someone who managed to figure things out without having any scientific tools at his disposal."

"Mist again?" I said. "I've missed that, Al."

"Here we go," he said.

When the air cleared and I could see again, I could make out two men sitting at a table outside what looked like a fine country home. There were vines on the nearby hills, and pine trees lining the avenue that led to the building.

"This looks like Italy," I said.

"Well spotted," Al said. "We're some distance outside Rome, and we're with Lucretius, a thinker and poet, who lived from around 99 BCE to 55 BCE. He's the one with the gray beard. The younger man is his scribe and translator, Caelius."

"Well, Caelius," Lucretius said to his assistant, "you've had time to go over my scrolls. What are your impressions?"

"I have to say that I'm worried," Caelius replied. "My family were not keen on me taking this assignment because you're known as being godless. That really comes across strongly in Book One of your poem. Are you not afraid of what the deities might do to you? And to me by association?"

"What poem is this?" I asked, looking at Al.

"And you think you're an educated man?" Al sniffed. "It's the famous *De Rerum Natura.*"

"That's Latin for *about the nature of things*," I said.

"Well done," Al said. "Lucretius wrote that poem to summarize his entire philosophy. Let's listen."

"Afraid of the deities?" Lucretius said, laughing. "Tell me, Caelius, why do so many gods appear and disappear throughout history? They do so faster than the ladies' fashions in nearby Rome."

Caelius frowned. "Perhaps it's because the true god has not revealed himself to us all."

"And why would he not show himself?" Lucretius said. "Why would he permit a constant parade of impostor gods to confuse us, his creations?"

"Maybe because we're not worthy of him?" Caelius said. "Or we've offended him in some way?"

"Oh dear, Caelius, you're starting to sound like those miserable whiners in Judea. Always moaning that their Yahweh god has forsaken them. Serves them right, that's all I can say."

"Very well, master, let me go over some things from your poem. First of all, the title. De Rerum Natura? What things are like? It doesn't exactly sound like a bestseller, does it?"

Lucretius smiled at the younger man. "It is what it is. I kind of like it. Next comment, please."

"Okay, sire. I must say I like the way you've made the preface sound like a hymn, and how you pay tribute to Venus as the mother of Rome. That'll help draw in some readers, before you hit them with your criticism of religious superstition.

"Then, you go on to talk about where things came from. I found that to be quite confusing. I mean, how you say that nothing emerges from nothing, and that nothing returns into nothingness. I don't get that."

"I see," Lucretius said. "I do write that the cosmos is made up atomies, countless number of them, residing in an infinite empty space. In doing so, I'm thinking back to the work of men like Democritus."

"The man who tried to split wood into its smallest parts?" Caelius said.

"The same. I speculate on how these atomies work, how they move and interact. It explains everything about our world and its contents."

"That takes us to the third book, master," Caelius said. "You make a case for man's body and soul being indistinguishable. When the body dies, so too does the soul."

"Therefore, fear of death and of misery after death is pointless," Lucretius said. "I hope you find that a comfort, Caelius."

"If you say so, sire," Caelius said. "As for the fourth and fifth books, I found them quite clear. The sixth does not appear to be complete as yet?"

"I still have some additional thoughts to record, Caelius. Perhaps you could write them down for me tomorrow morning? I confess to feeling somewhat unwell right now. I think I'll retire into the villa and get some sleep."

"Let me walk with you, master," Caelius said, casting a worried look at the older man.

"What's wrong with Lucretius?" I said, turning to Al as the images faded and we found ourselves back at the seashore.

"He's about to die," Al said, frowning. "He won't finish his grand poem. Sad to think that we lost the ideas still going through his fine mind."

"I'm impressed," I said. "I will definitely read his poem now. I like the way he managed to figure out the chaos of the cosmos without having any scientific tools at his disposal. We saw that earlier, when you showed me men like Eratosthenes. Brilliant, just brilliant."

"What do you plan to do now?" Al asked.

"I think it's time I pushed all my papers and books aside," I replied. "I need to spend some time thinking through what I've learned from all our trips through time."

"Very well," Al said. "But before you do, a word of caution."

HOTEL CALIFORNIA

The truth knocks on the door and you say, "Go away, I'm looking for the truth." And so it goes away.

—Robert M. Pirsig

CAUTION?" I SAID. "WHAT'S THAT ABOUT?" I looked at Al, wondering what was on the great man's mind.

"It's well known," he said, "or at least it's often said that physicists who get too near to the truth can go crazy."

"You mean like Boltzmann? I said. "We discussed him before."

"Sort of," Al replied. "Although in his case, I think he killed himself because he was stuck. He just couldn't solve all the equations cluttering up his mind."

"I see," I said. "You're giving me a warning then? You must therefore think that I'm close to a true Theory of Everything. I should take that as a compliment, shouldn't I?"

"Perhaps not," Al said. "You're not in the same league as the great ones who've gone bananas. In your case, the end game might just be that you start hallucinating a bit, or something harmless like that. There is that story about Hale, who had an imaginary elf acting as his assistant, although that might just be a rumor."

"Didn't Newton himself go mad?" I asked.

"Just manic depressive, I think."

"Maybe it's just that the best physicists get saddened by how little the rest of us know," I said. "We must look like fools wandering in the dark." I smiled at Al. "You know, Al, maybe the book I'm going to write will help humanity understand the truth."

"Hmm, yes, but about that caution I said I'd give you?" Al said. "Going deeply into the origins of the cosmos could be a bit like visiting Hotel California. You can check in any time you like, but you can never leave."

"If I go crazy, there's no way back?" I mumbled.

"You have been adequately warned," Al said, smiling at me. "Now I will leave you to your deliberations and will catch up you with you later. That's if you haven't hanged yourself."

With that, Al was gone. I immediately wondered if I hadn't already gone insane. After all, I kept having these visions of being guided by the world's most famous scientist.

Then I realized that, if I've already gone over the event horizon of my insanity, that there was no turning back. It was time for deep thought.

I knew what I had to do.

I left the seashore, went home, and made a pitcher of Midori Margaritas. Yes, this magic elixir will help release the truth knocking around in the canyons of my mind.

CONCLUSIONS

What did I learn?

CAPITAL LETTER

The present letter is a very long one, simply because I had no leisure to make it shorter.

—Blaise Pascal

"I HAVE WRITTEN YOU A LETTER, ALBERT," I said, smiling at him as he relaxed beside me, on the terrace at my home. I had laid out a selection of olives and chips to go with our Manhattans.

"Let me see it," he said. "I need proof that you actually know how to write."

I ignored the insult then pushed the pages across the table.

"I've finally developed my ToE—my theory of everything," I said. "I based it on all that I learned by listening to you and the people in history that we visited together."

Al leaned in and began to read. After a couple of minutes' silence, he looked across at me.

"Your handwriting is atrocious," he said. "Still, talk me through what's in here. Your thoughts on a creator, for example."

"Of course, Al," I said. "The first point in my letter is a recap of what we discussed in our earliest meeting. Namely, if you belong to a religion that believes in a divine creator, you have nothing whatsoever to puzzle out about the cosmos."

"Why not?"

"Because everything is the way it is because He—or She of course—wanted things this way. Grass is green, the sky is blue, rocks roll downhill, and waves crash on the shore—all because God likes that particular arrangement."

"You don't buy that, though," Albert said.

"I don't. Therefore, along with my fellow atheists, I have to explain where everything came from."

"The most awkward question of all," Albert said, smiling at me.

"However," I said, "the non-existence of a deity leads me to a couple of key thoughts. The first being—if there is no creator, then nothing can ever have been created."

"What do you mean by that, Ian?"

"I mean that there was no agent, no first cause of any kind, to design and build anything with any structure whatsoever. This then leads me to conclude that what happened had to be of the utmost simplicity. Not just in the origin of the cosmos, but in its subsequent development all the way to the present."

"Simplicity?" Albert said. "Then how do you explain what appears to be a rather complicated universe?"

"In due course, Al. First, let's consider the origin. Many believe that there was a moment when all matter in the universe was concentrated in a single focal point of unimaginable density, right?"

"Lemaître's cosmic egg," Albert said.

"He was quite wrong," I said. "I don't accept that there was any kind of material at the origin. No egg, and no cosmic dust, as some others claim. If there was such an egg, you'd then have to explain where that came from. So, there was nothing at all."

"You dare to contradict Lemaître? A man whom you agreed was superbly educated, having studied at Cambridge and Harvard?"

"I do, Albert. Perhaps you don't know this, but my clan motto is *I Dar*, an old form of the words *I Dare*."

"That doesn't mean a darned thing, Ian."

"I realize that. By the way, I also disagree with your own earlier belief that the cosmos was in a steady state. You were wrong about that."

Al looked sheepish. "Yes, well, it seemed right at the time, and then Hubble came along. So how do *you* say the universe originated?"

"Out of nothing, Al."

"Not a new idea, Ian. The ancients called that *Creatio ex Nihilo*. But you'll have to explain how it could happen."

"Challenge accepted. Here's why I think that the universe originated out of nothing. You see, it is an absolute impossibility for there to be a state where there is simply nothing."

"What do you mean by that?" Al said.

"Try to imagine it for yourself, Al. What would nothingness look it? No matter how long you spend thinking about it, you will eventually realize that it simply cannot be. Your mind desperately needs to fill the void. And that's what happens. The void is instantly filled."

Al nodded, then continued reading my letter. "So, because of that, a field of energy burst into being. You also say again that it had to be simple in structure."

"It had no structure at all, Al. Remember, it had to be of the simplest nature possible. Just a tearing apart of waves of energy."

Al looked at me. "So, a field or fields of energy—positive and negative charges ripping apart. You've also written here that the field has a net value of zero. Why?"

"Because," I said, "just as there cannot be nothing, there cannot magically be something either. That's why I suspect that the energy of the entire universe probably adds up to zero."

"How do we get planets and people out of that then?"

I looked at Al. "I think that charges rip apart evenly, into positives and negatives, except that some tear unevenly, while still having a total net value of zero."

Al chuckled. “The kind of thing we see in the Standard Model. Charges of one third here, and two thirds there, with their mirror images. But why would such uneven charges appear?”

I smiled. “Because nothing in the universe is perfect. So, the fields burst apart into unlike charges, pouring out into the cosmos. In fact, *creating* the cosmos as the chaotic energy pushes outwards.”

Al leaned in on the letter again. “It says here that this outpouring continues to this day, and might continue indefinitely? The universe would therefore be infinite?”

“I suspect so, Al, although I’m not completely sure about that. Some think that the acceleration and expansion we’re seeing will eventually ease off, like the eventual relaxation of a coiled spring being released. I prefer to think that if there’s nothing to stop the expansion, then why wouldn’t the cosmos continue to burst outward to infinity?”

Al nodded. “As you’ve written here, we know from Hubble’s work that the rate of expansion is accelerating. At least, for now.”

Al looked up and spent a few moments watching a woodpecker’s jerky movements as it climbed one of the nearby birch trees.

I interrupted his daydream. “My thoughts have another implication, Al, namely that the cosmos is a one-and-done event. No oscillating universe, I’m afraid, with everything collapsing and repeating time after time. Simplicity is again the watchword.”

He looked at me. “You’ve written here that the initial violence made it possible for us all to exist?”

“Al, if the initial outpouring had been even and smooth, we would have no structures in the cosmos. No elements, no solar systems, no life. All would be dark and quiet. But because there was no constraint on the initial explosion, it was violent beyond belief.

“That, in turn, created anomalies that became the nuclei of simple elements, like hydrogen. But the presence of such elements has no significance whatsoever. They can perhaps be regarded as meaningless space debris, or cosmic clutter."

"How do you explain the variety of the elements?"

"The core of an atom, the protons and neutrons, are held together by a strong force. But that force is localized, so the electrons whizzing around in their orbits are easily dislodged, and are free to make up other combinations, which in turn give us the denser elements. Those arrangements are created where there is intense pressure and heat, in the core of stars born out of swirling nebulae."

"Why do the nebulae swirl?" Al asked.

"Because the positive and negative charges of the energy waves behave like dipoles. We all know what happens when iron filings are affected by magnets—they spin into specific patterns. Everything in our cosmos is swirling in this way."

"Then why isn't the moon spinning?" Albert asked.

"Because over time the spinning energy fields of the sun, the earth and the ether have captured the moon in such a way that the forces match."

"It sounds like you are not a fan of gravity," Al said.

"Correct. I find the idea of a body pulling another from a distance to be ridiculous. Objects are where they are because they're being pushed. Everything in our cosmos is pushing everything else."

Al looked down at the letter. "Again, Ian, nothing new in this at all. This theory is popular with some. Although I find it a bit naïve."

"The good thing about being naïve, Al, is that there is always room for improvement." I smiled. "Let me continue. We were talking about the formation of the denser elements. When those stars burst out as supernovae, a huge assortment of elements was scattered into space, where everything could regroup in more benign locations, as planets like ours took shape.

"On the next page, I pick up on what so many believe to have happened. The pouring of solar energy into worlds like the Earth continued to excite the material collected there. Where circumstance favored it, amino acids formed, along with the other building blocks of what we call life. Individual components, such as sugar phosphates, connected to each other, not because anyone designed it so, but because they had a chance affinity for these connections."

"You write here," Al said, looking down at the pages, "that replication was also an accident?"

"Of course. As we discussed earlier, the material in DNA separates at points of weakness along the chain, perhaps simply because the chains get too cumbersome. There is, therefore, no purpose in any of the processes that led to life.

"Then there is the possibility of panspermia—namely that life could be quite common throughout the universe, and may even have come to our planet from elsewhere. It's possible, but it doesn't really matter whether it happened that way or not.

"What was important for our own history was the resultant long, gradual emergence of life in all its forms. We happened to end up with what we see around us every day: animals, trees, birds and bees."

Albert was reading on. "You seem convinced that there is no difference between life and non-life. I must say I sympathize with that view."

I nodded. "The boulder we talked about earlier? A tree, a horse, a human—those are just different arrangements of the underlying debris that the energy explosion has created."

"I see here that you tackle my issue of space-time relativity? You think you know why objects moving through space age more slowly?"

"I took a stab at that, Albert. I think back to what you told me about that lady in red. You know, the one from my dream?"

Al laughed. “Right, the fancy bar with a supermodel who you thought was lusting after you? Then she walked over to another, more handsome man?”

“Thanks for reminding me, Al. What you told me was that her body was made up of trillions of atoms and their fundamental particles or waves. I started to think of her as a huge cloud of buzzing, throbbing energy. When she started to walk across the room, she was in motion and therefore aging more slowly than I was. Do you agree?”

“Of course,” Al said.

“Perhaps because once her buzzing and throbbing cloud of energy goes into motion, it meets the huge buzzing and throbbing energy that makes up the entire cosmos. In other words, her cloud meets resistance from the other and more massive cosmic cloud.

“If I were running somewhere, Al, and I suddenly ran into a river, my speed would instantly be reduced, as my body had to fight its way through the deep water. I think of the lady in red in the same way.”

Al nodded. “The friction slows down her particular collection of atoms. They become slightly less active, so her metabolism is reduced.”

“Her cells replicate more slowly, so she ages less quickly than I do, still motionless at my table as she glides past me.”

“Her entropy is delayed,” Al said.

“As for entropy,” I said, “I learned a lot from your description of the two wire cages and the ping pong balls bouncing from one to the other.”

Albert smiled. “What would we do without ping pong balls?”

“The energy of the cosmos is constantly moving, Al,” I said. “It’s like a colossal version of your wire cages, with everything pushed and jostled around. As that happens, there is nothing to prevent a complicated structure from briefly taking shape and replicating—a flower, or a human—but as the shaking and vibrating continues, the tendency is for everything to degrade."

"I agree," Al said, nodding. "It's why a cup of hot coffee will calm down to become cold, but the reverse doesn't happen. What about time?" he said. "Did you tackle that?"

"You haven't reached that part yet, Al. It's on the last page. I believe that there's no such thing as time. There is no river of time, as Eddington claimed, or any kind of force carrying everything forward. There is only the constant motion of the energy in the universe. That's what moves the hands of a clock round and round.

"So, when someone asks me what existed before the cosmos emerged, I can explain to them that there was no *before*. We are all simply living in a massive cloud of surging motion that appeared because there cannot be nothing. What looks like the passage of time and aging, is just the rearranging of our personal clouds of atoms."

Al read the rest of my letter. "You describe us all as being tiny assortments of cosmic energy, twizzled by chance into what we call men or women. That the universe could just as easily have been smooth and even, without beings like us?"

"Exactly, Al. In other words, each of us is . . . a momentary lapse of simplicity."

There was a brief silence, then I smiled at him.

"I think I'll use that as my book's title," I said. "The idea of *A Momentary Lapse of Simplicity,* I mean. I like it."

"Hmm, as good as any, I suppose," Albert mumbled.

"What do you think, Al? Did my letter convince you?"

"It has some merit," he said, nodding gently. "There are, however, hardly any new ideas in this document you've written. Still I have to admit, grudgingly as it happens, that you have nicely summed up a lot of what many have come to believe occurred in the formation of our cosmos."

"Praise indeed," I said, pausing to smile at him. "In any event," I said, "I now I have my ToE—my very own Theory of Everything."

"Hang on," said Einstein. "What experiments are you going to devise, to test and prove out your hypotheses?"

"Experiments?" I said, somewhat confused. "We don't need no stinkin' experiments. I'll leave that for others who may be motivated to do that."

"So, you have no intentions of doing any real work whatsoever?" he asked, his face showing annoyance.

"Of course not," I answered. "Throughout my entire life, I have studiously avoided doing work of any kind. Remember why I did this, Al—I was trying to figure out the puzzle that came out of my conclusion that religion is a nonsense. That was—if there's no creator, then where did everything come from?

"As it happens, I think I've answered that quite adequately. To my own level of satisfaction, at least. All I needed for my selfish purposes was to develop at least one rational explanation—one without the need for any supernatural involvement whatever And I have done just that."

I THINK I THINK

One cannot conceive anything so strange and so implausible that it has not already been said by one philosopher or another.

—René Descartes

THERE IS A TEENSY PROBLEM or two with the theory you've developed," Al said, smiling at me. "Or, should I say, the theory that you've assembled from the work of others."

"Problems? Do tell."

"If you're correct, Ian, then planets and people are just little clusters of energy, twizzled by chance into the shapes that we see around us."

I nodded. "Yep, that's right."

"You've missed something," he said. "A couple of things, in fact. You see, I think there are many people who would follow along with the ideas in your letter, and they might nod in agreement."

"That's good, right?" I said.

"Yes, but they'd reflect for a minute or two, and many might then say it can't be true."

"Why not?"

"Because they have self-awareness," Al said. "Most people think that there is something more inside them than just neurons and veins and muscles. They think they have something more . . ."

"Metaphysical?" I asked.

"Exactly. It brings up the whole idea discussed by Descartes. He struggled with this issue for years, wondering whether he even existed at all. He ended up making a rather famous statement."

"*Je pense, donc je suis*," I said. "I think, therefore I am."

"Right. He ended up of the opinion that, since he could think, he did in fact exist, and was something more than just a dream."

Al looked across the patio table at me. "So, what would you say to René? And we could go back and visit him if you like?"

"Not necessary, Al," I said. "I think that our *ami* was too fascinated with his own thoughts. For me, the fact that I can think is not particularly special. After all, dogs, cats, and butterflies all think, to one degree or another."

"What about self-awareness?" Al said. "Isn't that more special than animal thinking? The way that we can perceive ourselves, and do the very kind of thinking that you've been up to in your deliberations? Isn't that exceptional in some way?"

"Nope. Sorry to disappoint, Al. It's still just neurons firing in patterns that came to have some usefulness to us in our human evolution."

Al was silent for a moment, then smiled. "There's another, related aspect of this. It's genius."

"My work shows I'm a genius?" I said, feigning a modest smile.

"Not in fact," Al said. "What I meant was that readers of your work will ask you awkward questions like, What about Mozart, or Bach?"

"Or Einstein?" I said, smiling at him.

"They'll say," Al went on, "that those talents simply cannot be the pulsing of cosmic energy, like the kind that drives plants and puppies. You have to admit, Ian, that Mozart's music is absolutely sublime."

"Not really. Too many notes, most of the time. I much prefer Mahler, or Sibelius. But, however sublime such music seems to us, it is of no consequence whatsoever to the cosmos."

"How do you make that out?"

I looked at him, then took another sip of my drink. "Picture this, Al. Imagine that we are hovering over a house somewhere in Europe. We look down, and somehow manage to speed up the passage of time. What would we see?"

"Not sure what you're getting at, Ian."

"As we speed up time faster and faster, the activity around that house becomes a blur. Babies being born, growing up, having their own families, but all of that shows up as a furious buzz."

"And so?" Al said.

"So, what if one of those babies turned out to be Bach? And then, we keep speeding on past that moment, with the centuries whipping by at breakneck speed. In that case, Al, what the heck difference has it made that one of those babies invented amusing little melodies that people reacted to?"

"That's very cynical, Ian."

"I think it's very accurate, Al. The cosmos is going to keep on expanding and, as the millions and billions of years pass, the human race will have died out, and the Bachs and Mozarts that we loved so much will be forgotten. Their genius, as we call it, is just the chance arrangement of energy that we happen to react to."

"You're depressing the heck out of me," Al said.

"Perhaps that's why some physicists decide to end it all," I said.

Al looked sadly at me. "But that's just plain awful, Ian."

"That's life, Albert."

NOBEL THOUGHTS

I intend to leave after my death a large fund for the promotion of the peace idea, but I am skeptical as to its results.

—Alfred Nobel

HELLO," I SAID, "IS THAT THE Royal Swedish Academy of Sciences?"

"It is, and my name is Ingrid," said a young-sounding voice on the phone line.

I thought she had a most attractive accent.

"What can I do for you, Sir?"

"I'd like to ask about the Nobel Prize for Physics. I believe your Academy is the organization that awards that particular honor?"

"We do," Ingrid said. "How can I help you with that?"

"It's early October," I said, "but I haven't received the call."

"The call?"

"Yes, Ingrid, to let me know that I've won the Nobel for Physics."

"I'm sorry," she said, "I think that one was announced earlier today, and it's gone to a Japanese researcher and one of her colleagues. Why did you expect to receive the prize?"

"I've just written a book," I said, "and in it I lay down an explanation for . . . well, pretty much everything. The origin of the cosmos and of life itself. Plus some pretty clever thoughts on time and entropy."

"I see," Ingrid said, her voice now sounding more hesitant.

"I also hear there's pretty decent money that comes with the prize?" I said. "Around a million bucks, right? I don't mind telling you, that would come in handy right now. I've got my eye on a new Cadillac, you see."

"This book you've published . . ." Ingrid said.

"Not quite published yet," I said. "It's drafted and I'm about to run it through the spell-checker. Should be out soon, though."

"I see," came the reply. "You know, Sir, a Nobel award is usually given for a lifetime's work in a particular field. It's also customary for the recipient to have been associated with prestigious academic organizations, and to have had his or her early work published and widely acclaimed. I'm wondering if your book quite falls into that category."

"I understand," I said. "It's just that I recall Obama got the Nobel Peace Prize before he did anything at all. I thought that meant I could sneak onto the award list."

"Sir, the reason that President Obama received the award was in recognition of the certainty that his mere existence as President was going to promote peace throughout the world. He was, in a certain sense, a Messiah."

"Okay," I said, "what about Literature, then? If I can't get Physics, could I get the Nobel for my book as a literary work?"

"That's handled by a different group," Ingrid said. "Namely the Swedish Academy. You could try calling them, and I can find their phone number for you if you like. I have to say, however, that you'll probably have no luck there either."

"How about Humor, then?" I offered. "Is there a Nobel for that?"

There was a long silence, then Ingrid spoke in a quiet tone.

"You know, Sir, our phone systems here in Sweden often have problems with dropped calls, and I sense that you and I are about to experience that any sec—"

It was quiet for some time, until I realized that Ingrid had gone.

How clever the Swedes are, I thought. Not only do they make great flat-pack furniture, and have that wonderful singing group Abba, but they can foresee when phone calls are going to be interrupted by transatlantic telecommunications glitches.

What to do now, I wondered.

I quickly had another idea, then used my computer to find the phone number I needed.

"Hello?" I said, when my call was answered. "Is that Columbia University in New York? I believe you administer the Pulitzer prizes. And I was just wondering how much money is involved?"

ONCE MORE INTO THE MIST?

Life is too important to be taken seriously.

—Oscar Wilde

"NO NOBEL PRIZE, THEN," Albert said. He looked sympathetically at me. "When I won mine, I gave the money away, you know."

"I read that somewhere," I said. "To your divorced wife, right? When I heard that, I realized that it must be true—physicists really are mad."

"Maybe so," Albert sighed. "It seemed like a sensible idea at the time."

"The folks who handle the Pulitzers weren't much help either," I said. "In fact, they seemed quite upset with me. Not at all like that nice Ingrid lady."

"Forget awards," Albert said. "They're all political anyway. Look, you may think you've solved the riddles of the cosmos, but that was just child's play. You have a much more difficult issue to deal with, and you're going to need help from the Time Lords again."

"A more difficult issue? What could be trickier than figuring out where everything came from?"

Albert chuckled. "It goes back to your original concern. You stopped believing in a god, right? That prompted you to figure out where everything came from, and you did okay with that. Well, now you have a bigger problem. Don't you see it yet?"

"Sorry, Professor, I don't."

"If there's no god, then whose rules are humans supposed to follow?"

I frowned. "Rules?"

"Yes," Albert said. "No god, ergo no Mosaic law, no Ten commandments. How on earth are people supposed to know what's right and wrong? How are we all supposed to live together?"

"I see," I mumbled. "Well, I suppose I could take a look at that."

"Time for mist," Albert said. "You are going back in time again."

"Oh good," I said, "I enjoy traveling with you."

"It won't be me," Al said. "I'm handing you over to another Time Lord."

"Who?"

"You'll see when the mist clears," he said.

THE END

OTHER WORKS

I find television very educating. Every time somebody turns on the set, I go into the other room and read a book.

—Groucho Marx

WE THANK YOU FOR READING THIS BOOK, and we do hope that you enjoyed it.

To be informed of new works by Ian G. Dalziel, please send an email to *info@TheDLPages.com.* Your email address will only be used to inform you about our books. It will not be disclosed to third parties.

The author's works to date have covered several fields of interest. (Yes, he's a brilliant novelist too.) Here is a list of his published works:

The Curious Evolution of Christianity (Non-Fiction): The author was brought up as a Christian, but stepped away from that religion when he had too many doubts about what he was hearing in church. This inventive and amusing book is the result of his research into what really happened as that religion took shape over three millennia.

Named for Snow (Fiction): Stassi works hard at her art, hoping to sell her paintings in Manhattan galleries. By chance, she takes down a small ad from a notice board, then finds herself in touch with a mysterious and elderly Japanese man, an art collector. He and his granddaughter are about to change Stassi's life in ways she could never have expected.

Princess of the Allevanti (Fiction): In an ancient time, a beautiful princess is captured, then forced to scan the night skies looking for omens. When she sees the one her captors fear the most, she is taken to meet their King.

She Dyed for Love (Fiction): A romance in the old Hollywood style Linda is a top movie star and seems to have the perfect life. So, why does she go into a beauty salon and ask them to turn her into a blonde?

The Kashmir Shrine (Fiction): A thriller with an unholy twist. How two young women try to solve the mystery of an old man's work with religious relics, and his experiments with ancient DNA.

Contact in Nevada (Science Fiction): An archeologist stumbles on a way to contact aliens, and becomes involved in a race to prevent the world from widespread war.

Prostatectomy (Health): The author's informative and at times witty, account of his own battle with prostate cancer. Helps men diagnosed with the disease to know what may lie ahead.